THE AVERAGE BOOK

THE AVERAGE BOOK

by Richard Smith & Linda Moore
illustrated by Edie Bowers

The Rutledge Press
112 Madison Avenue
New York, New York 10016

ACKNOWLEDGEMENTS

A special thanks to Kathy Williams for her "two-handed" editing, a pencil in one and a calculator in the other. Also an above average thanks to both Noni and Brynn.

Graphics by A Good Thing, Inc., New York, New York
Composition by Sandcastle, New Rochelle, New York

Produced and Packaged by David M. Cohn Publishing, Inc.

Published by The Rutledge Press

Distributed by W. H. Smith Publishers, Inc., 575 Lexington Avenue, New York, N.Y. 10022

First printing 1981
Printed in the United States of America

ISBN: 0-8317-0593-0

Contents

Introduction

Throw a bunch of numbers together and divide by how many there are and you have an average; so the world is full of averages. But people find most averages dull and couldn't be bothered reading a book full of them. So we've chopped out the dull ones and categorized the rest in an effort to fascinate, titilate and up-date you.

Many of our averages have to do with your money: how much you earn, how you spend it, how you compare with the mythical "Average American;" or in some categories, with people around the world. Other averages allow you to compare your city or state with others, or the U.S. with foreign countries. Probably the most enjoyable averages are those that reveal people's personal habits: eating patterns, family relationships, sexual practices, religious activities, personal hygiene, patriotic feelings and so on. The averages we've chosen have a lot to say about the state of our society and our world, and your own place in them.

Discover how the other half lives. Read on!

1

The Human Body

The average number of Americans that report that someone in their family has suffered from mental health problems severe enough to affect their physical health is 1 out of every 3.

Back talk –
The number of Americans with severe back problems averages 1 out of every 4.

Americans spend an average of $5 billion annually diagnosing and treating back problems.

On the average, a person would have to eat about 11 pounds of potatoes to gain 1 pound of body weight.

The average person grows a new eyelash every 3 months.

On the average, hair grows faster in the morning than at any other time during the day.

Go east, young man! Stateside, women wear an average bra size of 34B while in Europe the average is 34C.

Following a corneal transplant operation, the average length of time during which driving is restricted is 6 weeks.

The life span of a caveman averaged only 19 years.

The average cost of having a fact-lift done using acupuncture is $3,500 for the 9 treatments involved.

A recent 9-year study indicates that an average of 31% of prescribed hysterectomies were not deemed necessary when the opinion of a second physician was sought. They also found that 29% of bunions recommended to be removed surgically, 27% of knee surgery, 16% of breast surgery, and 15% of cataract removals were not suggested by a consulting physician.

Females average more dental visits per year than do males (1.7 for women, 1.4 for men).

The average lactating mother produces $1\frac{1}{2}$ pints of milk a day.

The average left-handed person tends to be more creative and receptive to new ideas than the right-handers, a study shows.

People dream an average of 5 times a night, and the dreams keep getting longer. The first dream is about 10 minutes long and the last is about 45.

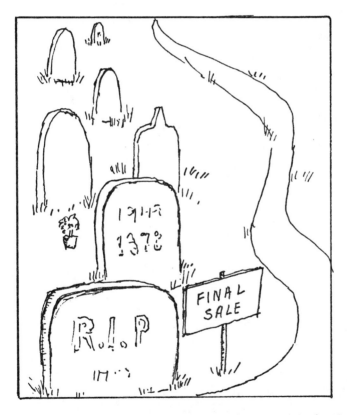

An average of 2 million people die annually in the U.S.

On the average, women dream more than men and children dream more than adults; more people dream in black-and-white than in color.

On the average, 1 out of every 50 Americans is manic-depressive.

The average human male produces 138,000 sperm cells in 60 seconds.

The human head sheds an average of 68 hairs during a 24-hour period.

Tailors have always known than in an average of about 85% of the male population the left testicle is larger than the right.

On the average, women have a slightly higher IQ than do men.

The average length of time it takes a worker's body to adjust to a new work shift is 21 days.

Dr. Michael DeBakey reports that of 1500 coronary by-pass patients, the number still working 5 years after surgery averages 4 out of every 5 of those younger than 65.

Ready to cash in your chips?
An anatomy professor recently calculated that inflation has driven up the value of the minerals in the average human body from 98¢ in 1970 to $7.28 in 1980.

On the average, more than half of the people bitten by poisonous snakes in the U.S. are never treated and still survive.

Weighing in at 11 ounces, the average heart pumps 4 liters per minute.

The average life expectancy of a woman in Gabon, Africa, is 45 years, but her male compatriot will live only 25 years.

The average Japanese person has 10 times more mercury in his body than does an American or European, due to the high percentage of fish in his diet.

The average number of people dying from choking on food each year is greater than the average number of people killed by guns, airplane accidents, snakebites, or electrical shocks.

The average brain weighs only 3 pounds but contains 10 billion nerve cells, each capable of 25,000 interconnections.

The human tongue contains an average of 9,000 taste buds.

The chances of men getting leprosy average 2 to 1 over women contracting the disease.

Babies born in May weigh an average of 200 grams more than babies born in any other month.

On the average, 5 million Americans are hospitalized annually for stomach ailments other than those considered to be diseases.

On the average, 1 brow wrinkle is the result of 200,000 frowns.

On the average, 1 out of every 5 American men and 1 out of every 4 American women are at least 10% overweight.

On the average, a teenager is 50% more susceptible to colds than people over 50.

The average American is eating fewer calories than ever, but average weights have not fallen since 1965. While Americans are eating less protein, fats, and carbohydrates, they have increased their consumption of sweets, soft drinks, and alcohol tremendously.

The average heart beats 100,000 times every 24 hours.

The average college-age man is about 15% body fat. A woman the same age averages 25% body fat.

On the average, 2 out of every 5 American women dye their hair.

The average number of teenagers suffering from acne is 9 out of every 10.

Despite health warnings, the average American still eats food with preservatives and coloring; 30% work long hours and smoke cigarettes; and 20% are overweight.

Jaws–
The average child grows his full set of permanent teeth by age 17.

The average American spends about $600 each year on medical expenses.

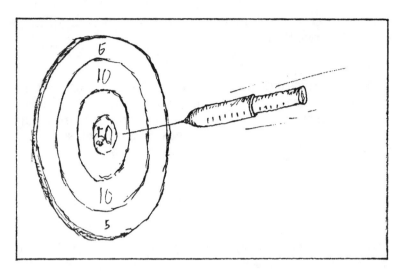

Good shot!
A tetanus shot remains effective for an average of 10 years.

On the average, 47% of Americans' medical expenses go to hospital costs, 19% to doctors, 10% to drugs, 8% on nursing homes, and 7% on dental expenses.

The average spinal cord is $\frac{1}{2}$ inch thick.

Germ watch-
The average American hospital patient has a 5% chance of picking up extra germs during a hospital confinement.

And the beat goes on -
At rest, an average baby's heart beats 130 times per minute; a 10-year-old, 90 beats per minute; an adult female's, 78 beats; and an adult male's, about 70 beats.

And not a minute sooner –
On the average, Canadians are willing to donate a vital organ at death, although the percentage drops as the age increases.

An average Watusi female measures 5 feet 8 inches when mature.

Hair today, gone tomorrow–
The average hair grows about ½ inch per month, or .33 mm per day. Furthermore, a single hair can hold a weight of 3 ounces.

On the average, a woman is 3 times more sensitive than a man to noises while sleeping.

The average person could survive 6 days without water, assuming that the outdoor temperature was around 60 degrees.

Fifty lashes!
Each human eyelash lives an average of 150 days.

In 1 square inch of skin, there is an average of 645 sweat glands, 63 hairs, 18 feet of blood vessels, and 78 yards of nerves.

An Army study shows that, on the average, 1 out of every 4 members of the U.S. armed forces is overweight.

The average age for breast development is 12.4.

The average 14th century European could expect to live to age 38. Three centuries later, life expectancy averaged 51 years.

The average female between the ages of 20 and 44 is more likely to be overweight than are males in that same age category.

On the average, black Americans are twice as likely as white Americans to contract diabetes, a fact as yet unexplained.

The average American can remember 5 to 9 numbers in correct order.

The average age of Americans in 1980 was 29.9 years. Population experts predict that it will be 31.1 years by 1985, 32.3 by 1990, and 34.8 by 2000.

The heaviest organ in the human body is the liver, which averages $3\frac{1}{2}$ pounds.

The average human liver is more than 5 times the weight of the human heart.

Ancient Egyptians lived an average of 29 years.

On the average day, about 700,000 Americans are short-term hospital in-patients.

The average person flexes his or her finger joints 25 million times during a lifetime.

The incubation period for a case of leprosy averages 3 years.

Americans who wear eyeglasses average about 1 out of every 2.8 people, totaling about 100 million.

The average human kidney contains about 70 miles of tubes with millions of tiny filters. The total blood supply of the body is filtered through the kidneys in less than an hour.

An adult human hair can stretch an average of 25% of its length without breaking.

Gimme some skin-
The average adult has about 18 square feet of skin, weighing about 6 pounds.

Doctors say that on the average, 2 out of every 5 of the 3 million Americans who visit Mexico each year are stricken with the dreaded "Montezuma's Revenge." Taking 60 milligrams of Pepto-Bismol 4 times a day seems to cure the diarrhea.

The average person inhales 1 pint of air with each breath, and breathes 10 million times a year.

The average woman's life span is longest in Florida; men live longest in cooler Connecticut.

Sexual intercourse occurs most frequently during ages 21 to 25, on the average.

Patients from families earning less than $5,000 per year averaged 5.8 visits to the physician per year, contrasted with 4.7 visits for $10,000 income families.

The adult lungs contain an average of 300 million air cells.

Heart attacks that occur on the job strike their victims on Monday morning in an average of 3 out of every 4 cases.

Input and output-
The average person consumes from various sources about 3 quarts of water a day and passes about half of it as urine.

On the average, 4 out of every 5 migrane sufferers are women.

The body of an average-size fully-grown human male contains enough fat to make at least 7 cakes of soap, enough phosphorous to make 2,000 match heads, enough carbon to make 8,500 pencils, enough iron to make 1 nail, and enough water to fill a 12-gallon barrel.

There are about 12 billion cells in the average human brain.

Water constitutes an average of 73% of the human body. A lean man may average as much as 75% while a woman may have only 52% of her weight in the form of body water.

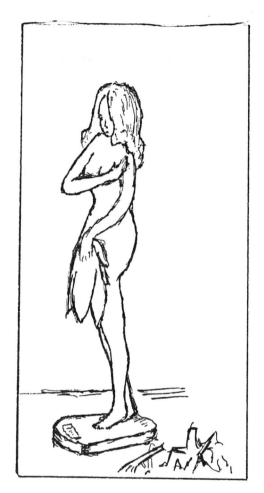

Take it off!
A stripper loses an average of 36 calories during her
15-minute act.

An average number of Americans on some form of diet
is about 1 out of every 2.

The average human brain triples in size between birth
and death.

The average length of time a person could survive the icy waters of the Arctic Circle is 1 hour.

The average adult consumes his own weight in food once every 50 days.

After you fall asleep, your first dream averages 10 minutes.

Nearly 1 out of every 2 males graduating from college in 1980 had a beard and/or a mustache.

Fingernails average faster growth in the summer than winter. The middle finger averages faster growth than the thumb.

On the average, you are more likely to catch a cold when the germs are passed from hand to hand rather than from mouth to mouth. More cold viruses appear on the hands than around the mouth.

Male drivers' reflexes average 1/10 of a second faster than females (.593 versus .667 of a second).

The homicide rate of black Americans averages 4 times that of black Africans.

Cincinnati puts the gentlest bite on dentistry patients in the U.S. The average dental bill is $15 per visit.

A blond beard grows faster, on the average, than a dark beard.

In a 1980 survey on stress, an average of 9 out of 12 men considered themselves "happy," while only 8 out of every 12 women felt happy.

Water on the brain-
The average brain consists of 85% water.

Of people recently surveyed, an average of almost 3 out of every 4 claimed to have had a thorough medical exam within the past 2 years; more than 1 in 2 claimed to have had one within the last year.

The average ratio of blood cells in a healthy body is 800 red to every 1 white cell.

The average complaint of patients not receiving adequate medical care is that they have trouble getting an appointment, rather than the medical costs being too high to afford.

After coming in contact with a rabid animal the average time span for symptoms to appear is 30 days.

The probability of 2 people with the exact same lens prescription for eyeglasses is 1 in 4 trillion!

High there-
At birth the average Causcasian boy measures 19.8 inches long, while the average male pygmy is 18.0 inches. By the time they are grown the difference will increase to 9 inches (5 feet 9 inches to 5 feet).

The average American man is 5'8" tall.

Thick skinned-
Skin on the soles of the feet average 1/6 of an inch thick while that of the eyelid averages only 1/15 of an inch.

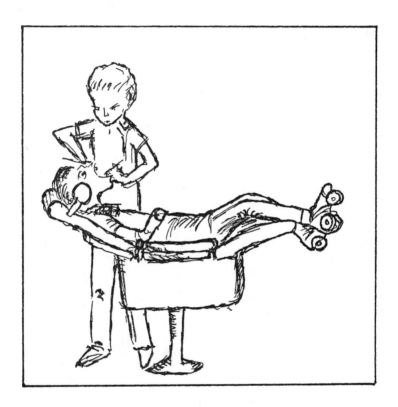

High rollers ...
According to the lastest figures available, roller skating injuries have more than doubled in the past 6 years. The average facial injuries include broken noses, broken jaws and knocked-out teeth.

The average pulse rate is 85 beats per minute.

Dubious distinction-
The African nations of Angola and Malawi average the
world's highest death rates.

The average adult head weighs 12 pounds and is made of
a solid bone.

So--Nu?
An Israeli uses an average of more than 200 hand and
facial gestures, compared to 65 for an American.

About 1.1 million Americans are long-term hospital
in-patients on an average day.

The average person can distinguish 150 different colors;
while the trained eye can detect over 100,000 different
shades and hues of color.

A recent study of drowning victims concluded that an
average of 4 out of 5 who are submerged for more than
5 minutes can be revived with resuscitation techniques.

Nuns have an averge life expectancy of 77 years, the
longest of any group in the U.S.

A snore can register at 58 decibels; the average
pnuematic drill is 82.

Hour upon the stage-
The average life expectancy at birth of the whole
human race is about 60 years.

By the time children reach the age of 2, tooth decay will have already struck 1 out of every 2 of them. By 17 years of age, the average number of teeth that have suffered tooth decay is 9.

Every cell in the human body is renewed on the average of every 7 years.

The average person spends about 40% of his waking time daydreaming.

Most daydreams average 14 seconds in length.

The average adult stomach is the size of a clenched fist.

The annual death rate of U.S. males from all diseases averages 888 per 100,000; for women, 727 per 100,000.

An average of 4 out of 5 members of the high-risk population have not been vaccinated against the flu.

An average step forward involves 54 muscles.

An average of more than 9 out of every 10 school-age children in the U.S. have been vaccinated against measles.

A survey has determined that an average of 1 out of 2 is hard of hearing and of these 1 out of 3 is considered partially deaf.

A University of Chicago researcher finds that babies under 2 years of age average 50% more interconnections in their brain cells than average grownups.

How can you win?
People with incomes over $50,000 a year live 2 years less than the average person.

About 2 out of every 3 Americans exercise on the average of 2 or 3 times a week, according to the Health Insurance Institute.

The average life expectancy of Chinese men and women is 50 years.

Career women are up to 30% more likely to outlive in average life expectancy non-working women, according to an insurance company study.

The risk is 4 times greater, on the average, of a child becoming an alcoholic if born to alcoholic parents, rather than to non-drinkers.

The average number of American children today who have high cholesterol levels is 1 out of every 3.

The average hospital stay for Northeasterners is more than 8 days, compared to less than 6 for Westerners.

According to a worldwide medical study Americans have the highest average number of headaches.

The path not taken-
The average brain size of a Neanderthal man was larger than that of the Cro-Magnon man, but the former became extinct and the latter survived to become our common ancestor.

It takes an average time of an hour and a half to embalm a corpse.

The Gillette Razor Company reports that the average man removes about 30,000 whiskers during the average 10-minute shave.

A man's beard grows an average of $5\frac{1}{2}$ inches in a year or 30 feet in a lifetime.

The average person who has a heart attack has a cholesterol level of 244 (244 milligrams per 100 milliliters of blood).

The average person's body makes 1 gram of cholesterol a day.

An average of 2 out of every 5 Americans have never visited a dentist.

About 1 out of every 20 Americans, on the average, suffers from phobias (unreasonable fears). Doctors now list over 700 different types of phobias.

An average of 1 out of every 6 people is left-handed.

The average dose of aspirin is absorbed into the bloodstream in 30 minutes.

Female's lips average more wrinkles than men's because women smile more.

The average face-lift lasts from 6 to 10 years, depending upon care and stress. The operation reduces a person's apparent age by about 10 years.

In an average set of human lungs, there are about 600,000 alveoli or small branches of bronchial tubes.

The average human stomach contains about 35 million digestive glands.

The human heart, beating about 100,000 times per day, pumps an averge of 13,000 quarts of blood daily.

The average man has 25 trillion red blood cells; the average woman, 17 trillion.

Wotta Schnozz!
An average professional "nose' or smeller in a French perfume factory can recognize more than 7,500 different odors.

The golden age?
The average lifespan of the Cro-Magnon man was 32 years; Neanderthal man, 29; the Ancient Greeks and Romans, 36.

An average of 1 out of every 40 Americans claims to be a vegetarian.

An average of 1 woman in 1,000 has color blindness; but 70 men out of 1,000 suffer some form of it.

It's going up down-under-
New Zealanders have an average life expectancy of 71 years.

Bedtime story-
On the average, 1/3 of our lives is spent sleeping, and 1/5 of our sleep spent dreaming.

Hair-raising-
New hairs grow at an average rate of 0.33 millimeters a day except when you are in love and hair growth speeds up.

Most newborn babies sleep an average of 19 hours a day. But 6 year olds sleep 11 hours; 12 year olds, 8 hours; 25 year olds, 8 hours; 40, 7.5 hours; 48, 6 hours; and at 60 or older, 5.5 hours.

The human heart averages at least 1 beat every second, performing work equal to lifting 70 pounds about 1 foot vertically. By the age 70 a person's heart has pumped at least 2,800,600,000 times.

The average range of the human voice is about 200 yards.

Lift off-
The following are the average cosmetic surgery costs:

>hair transplant $4,000
>face lift $3,500
>eyelid lift $2,000
>dermabrasion $1,500
>nose resculpting $2,000
>chin augumentation $750
>breast enlargement $2,500
>breast lift $2,500
>breast reconstruction $3,500
>breast reduction $3,500
>tummy tucks $3,500

An opera singer burns an average of more than 2 calories per minute during a performance.

Miles of files-
An average of 2 out of 3 adults in the U.S. have suffered from hemmorhoids.

The mature male body averages about 40% muscle; the female body averages 30%.

An average of 3 out of every 4 Americans visited a physician during the past year.

It takes an average of 147 days for a fingernail to grow from the cuticle to the tip.

Babies born during the month of May average 7 ounces heavier than babies born any other month of the year.

The average life span of human red blood corpuscles is 127 days.

An average of 9 out of every 10 suicide victims are females.

Untimely plucked–
A tweezed hair will reppear on the skin in an average of 57 days.

U.S. hospitals treat an average of 7 million patients a day.

The average human body replaces 15 million red blood cells each second.

An average of 1 out of every 4 Americans had surgery in the past 5 years. Among those, 1 out of every 3 got second opinions before proceeding with surgery.

The average man gets along on less sleep than an average woman can.

The average person begins to shrink in height after the age of 30.

The time of day with the highest average number of births is 4 a.m.

The average woman's thighs are $1\frac{1}{2}$ inches larger than the average man's.

The average human body looses 2.5 quarts of water a day; some as urine, some as perspiration, some as breath.

The average life expectancy of the American Indian is 46 years, 23 years below the national average.

The average snorer is likely to be an overweight man.

The human hand contains an average of 1,300 nerve endings per square inch!

The average newborn baby's head accounts for about a quarter of its total body weight.

The average child nees 8 to 9 hours of sleep a night to feel rested. The average elderly adult needs 4 to 6 hours.

John Hopkins magazine reports that on the average day a person laughs more than 20 times; cries once a month; and laughs an average of 400 times before crying once.

The average urbanite walks about 6 feet per second, twice as fast as his small-town cousin.

The U.S. Public Health Service reports an average of about 3 out of every 4 Americans age 20 or older use aspirin to relieve occasional aches or pains. People with higher education and income use aspirin significantly less than others.

The average human eye distinguises an average of 500 shades of gray.

A study showed that only 42% of sleeping people woke when a smoke alarm was sounded at intervals during the night. The average alarm is not loud enough to wake the average sleeper.

Psychological studies on reaction time show that when an average man's senses warn him of approaching danger, he reacts much faster than the average woman.

There are an average of 580 hairs per square inch in a man's beard.

Take 2, and call me in the morning-
Americans swallow 17 billion aspirin tablets per year; that's an average of 77 aspirins per person.

The average woman's heart beats faster than the average man's heart.

Gum disease affects an average of 9 out of every 10 Americans.

An average of 15 million red blood cells are produced and destroyed in the human body every second.

The average brain is 2% of a person's total body weight.

Human bones average 25% water, 45% mineral (mostly calcium), and 30% living tissue, such as blood vessels and cells.

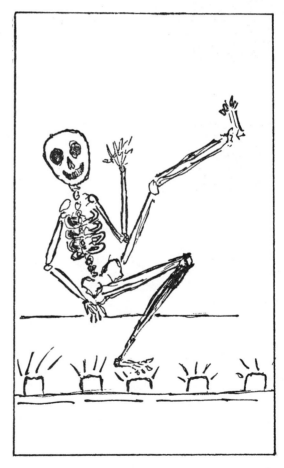

The average human heart beats 100,000 times per day or 2.5 billion times in an average lifetime of 72 years.

The average child-bearing age for women is in their 20s. The oldest woman to give birth was an American aged 57. The youngest mother was a Chilean girl only 5 at the time she gave birth to a live child by Caesarian section.

The average Romanian still indulges in mud-bathing. This ancient practice is said to help almost any ailment. The Romanians cover their bodies with mud and then bake in the sun.

2

Marriage and Sex

On the average, 2 out of every 3 of the women surveyed told <u>McCalls</u> magazine that emotional problems and inadequacies were the hardest subjects to talk about with husbands or lovers.

In a recent survey, an average of nearly 3 out of every 5 men said that the most unpleasant aspect of sex would be to make love to an "unresponsive" woman.

On the average, men who underwent sex-change operations and became women earned less money at their new careers than when they functioned as men.

The human sperm travels an average of 1/10 inch per second.

On the average, more than 3 out of every 4 American adults are married.

The average amount of energy used during sexual intercourse is comparable to the amount required to climb 2 flights of stairs.

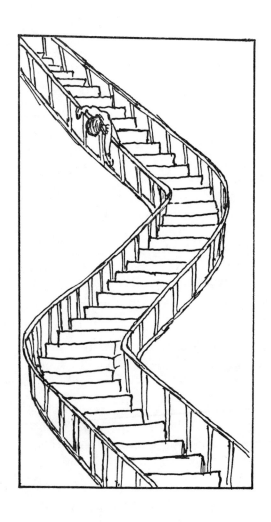

People polled who said that they would definitely marry the same mate again averaged 3 out of 4. Of the others, who said "no way", 1 out of every 4 was a woman.

Of every 10 men in the U.S., an average of 1 is impotent.

The average interval episodes of masturbation among 30-year-old males is 6 days.

The average age for first marriages is 23.7 years for men and 21.3 years for women.

The average number of divorced men and women who claim that extra-marital affairs caused their divorces is more than 1 out of every 2.

The average wedding feast in Yemen lasts 21 days.

The average under-30 person considers himself or herself romantic and thinks that women are more romantic than men.

High-income people ($18,000 a year or more) are, on the average, much more likely to consider themselves romantic.

What irritates the average male most during sex? Men surveryed responded: "cold or disinterested woman" (60.1%), "a woman who criticizes" (11.5%), and "woman who makes demands" (4.9%).

On the average, couples in their second marriage are twice as likely as divorced people to call themselves "happy."

The average frequency of intercourse for married couples between the ages of 35 and 44 is 98 times a year.

The average number of men surveyed that said they use the female-superior position of coitus at least occasionally is 3 out of every 4.

Just in time for the commercials!
The average married couple making love engage in foreplay for 15 minutes, followed by 10 minutes of intercourse.

Be my love-
According to the Roper Opinion Pollsters the average man is primarily attracted by a woman's sex appeal and the average woman is attracted to a man's sense of humor and his sensitivity to feelings of others. Both men and women rated intelligence high on their lists.

On the average of 3 out of every 4 American men do not feel that women should receive alimony if she is able to work.

The male chauvinist pig is disappearing-
The number of married men that ask their wives for career advice averages 4 out of every 5.

Position is everything-
According to researchers the average man would like to try more and different sexual positions.

Get that degree!
Reports show that the average number of male college graduates having success in bringing sexual partners to orgasm is 7 out of every 10; high school dropouts scored only a 4 out of 10 success rate.

Practice makes perfect-
According to a study conducted by M. Hunt, the number of people under age 25 that enjoy and practice oral sex averages 9 out of 10.

According to a recent survey, the average number of American men who experience sexual intercourse by age 17 is 9 out of 10.

When there's a will–
Sex surveyors report that more than half of the men questioned claim to bring their partners to orgasm by means other than intercourse, at least half the time.

A poll taken in 1980 indicates that an average of more than 12 out of every 15 women felt that they should be the chief decision makers in their families. When the poll was taken in 1962, only 5 out of 15 felt thay they should make the most important decisions.

Results of a sexual behavior survey from 1974 show that an average of 3 out of 4 single women had sexual intercourse before age 25.

According to a psychological study, an average of only 1 out of 5 recently divorced men felt regret over the breakup while nearly 1 in every 3 women felt sorry.

On the average, 1 out of every 3 jail inmates has never married.

A sex study finds that an average of 3 out of 5 women always achieve orgasm with 15 to 20 minutes of foreplay.

An average of only 2 out of 5 women reach orgasm consistently with less than 10 minutes of foreplay.

The average contraceptive diaphram size is 70-75 millimeters.

The divorce average in San Diego Superior Court is 30 marriages dissolved every $2\frac{1}{2}$ minutes.

On the average, according to 3 separate sex studies from 1948 through 1976, about 1 out of every 2 women achieves orgasm almost every time she had intercourse.

For an average of 4 out of 5 men, sex is not the most important pleasure in life, according to a recent study. Career success and love are considered more important.

The egg and I-
The average human male ejaculation contains 300,000,000 spermatozoa.

In a study of over a thousand average American husbands, over 50% do the shopping, 70% claim they can cook, 44% can prepare a complete meal, and 14% clean the oven.

The average length of time a woman remains a widow in the U.S. is 11 years.

Mr. Clean-
An average of 4 out of 5 men interviewed said they would not enjoy househusbanding if their wives were the breadwinners. Typical beg-offs were (a) don't like housework, (b) don't want to stay at home, and (c) not manly.

Divorce averages are highest during the month of July, followed by October, then June.

Some people have it all–
Researchers who rated people's attractiveness on a scale of 1 to 10 discovered that those who rated 3 had an average income of less than $10,000; those who rated 6 earned between $10,000 and $20,000; and people who received the highest rating earned more than $20,000 a year.

It's a gamble–
In 1975 the marriage mills of Nevada performed mariages at 17 times the national average of 10.1 per thousand population per year.

On the average, a second marriage is happier than the first.

An average of 4 out of 5 men surveyed claimed they tried to delay orgasm to increase their partner's pleasure.

The average number of American women that become widows is 3 out of every 4 and only 7 out of 10 will remarry.

An average of 7 out of 10 people in a recent survey admitted to being sexually aroused by erotica, such as pictures, drawings, music and/or prose.

The contraceptive technique with the highest average of perfection is total abstinence. The pill has the second best average. Estimates show it has failed only 0.1 to 2.0% of the time.

The average duration of morning erection in men is 40 minutes.

Catching up-
A poll of recently divorced men indicated that these men had an average of 8 intercourse partners in a year.

The average frequency of intercourse for married people 25 years and younger is once every 2 days.

Breakin' up the gang-
An average of almost 4 out of 5 single men surveyed expressed an intention to marry or remarry eventually.

In 1870 the average cost of a divorce in Utah was $2.50.

Chivalry is not dead-
An average of more than 1 out of 2 people do not consider a sex act over until their partners have experienced orgasm.

An average of 5 out of every 6 men earning $60,000 or more a year are married.

We-e-l-l, maybe-
An average of 1 out of every 3 men in a recent survey claimed they would never cheat on their regular partners.

The highest average frequency of masturbation for humans occurs aroung the age of 15. The maximum frequency exceeded 24 times a week.

Head in the clouds-
A research team studying the love lives of skyscraper workers found that 2 of every 10 people working on the lower floors were romantically linked with a co-worker. Yet on the upper floors, 9 out of 10 people became involved.

Honey, I'm --oops!
Somewhere between 2 and 4 seconds before he ejaculates, the average man knows it's going to happen. Usually the young man can ejaculate a distance of 12 to 24 inches. Age diminishes the distance to 3 to 12 inches.

The average duration of marriages that end in divorce is 7 years.

The impossible dream-
A sex survey found that the average 18 to 24-year-old male wanted to have intercourse 5 to 7 times a week but actually averages 3.5 times.

Habitual male masturbators report that they do it on the average of once every 2 weeks at age 60.

Among average men aged 55-65 the desire for intercourse is about twice a week, according to a recent sex survey.

American women favor the pill for contraception on an average of 3 to 1.

In the average year, approximately 250,000 men are attacked and beaten by their wives.

Only you-
An average of about 2 out of 3 men surveyed said that their mind stays on their current sex partner during intercourse.

An average of about 2 out of every 3 men questioned in a recent survey reported that they avoided intercourse during their partner's menstrual period.

It's fun anyway-
An average of 3 out of every 5 respondents in a recent survey said that they enjoyed hugging and kissing even when it did not lead to intercourse.

In a recent Roper poll, an average of 3 out of every 4 Americans questioned thought that homosexuality is "wrong."

The average price for a bridal gown was $300 in 1979.

The average cost of a wedding reception (an elegant sit-down affair at a downtown hotel) was $80 per guest in 1979.

For a home wedding reception, hor d'oeuvres, champagne, and cake averaged $16 per guest in 1979.

A wedding luncheon or brunch with several courses served averaged $21 per guest in 1979.

An average of nearly 1 out of 3 American Indians now marries outside his or her culture.

An average of 9 out of every 10 women polled said that they had been sexually harassed on their jobs.

The average American man falls in love 2.5 times in his life.

The average time from marriage to first child for women married between 1975 and 1978 was 2 years, 10 months longer than for women married between 1960 and 1964.

August, July, and June are the months averaging the most weddings.

An average of 4 out of every 5 women polled considered an unstable marriage a liability for a politician, but 96% of them would vote for someone who was divorced.

I do-
The average length of a marriage engagement is 300 days.

The average length of a love affair is 18 months.

An average of 7 out of every 10 men surveyed felt that love was the most important thing in life and that sex was much better with love.

The average American man will have 1.7 extramarital affairs during his lifetime.

The average American woman will have .6 extramartial affairs during her lifetime.

The average American woman falls in love 4.8 times in her life.

3

Family Life

During an average week, 40% of all Americans attend church, the highest percentage being among West-Central Catholics.

According to the General Telephone Company of Pennsylvania, the average American spends approximately 1 year (8,760 hours) of his or her life talking on the telephone.

Market research shows that an average of 3 out of every 4 consumers use coupons for supermarket shopping.

During the first 9 months of 1980, the average American spent $441 for clothes.

An average of more than 1 out of every 3 newborn babies is breast-fed, a figure up 40% since 1973.

A retired couple in an average U.S. city needs an annual income of $6,023 just to get by; $8,562 for moderate comfort; and $12,669 to live very well.

The average cost of keeping a baby in disposable diapers for 30 months is $1,100.

The average cost of washing and drying cloth diapers used over a 30-month period of time is $200 (including the detergent and bleach).

On the average, out of every 100 homes in America, 3 are without a bath or shower, 2 are without flush toilets, and 1 is without running water.

The average life span of an electric can opener is 10 years; electric knives and hot plates, 15 years; food processors, 8; corn poppers, 7; and coffee makers, 3.

The number of high school students taking French, German, or Russian beyond the second year averages only 1 out of every 20.

Americans drink an average of 12 gallons of tea per year. Almost 80% of that is iced tea.

Since 1970 college attendance by women has increased an average of 5.6% a year.

Homes built in the U.S. today use an average of 10% less energy than those built 5 years ago.

The average Englishman drinks 2,000 cups of tea each year; that's almost 6 cups per day!

It is expected that the average home in the year 2000 will use 50% less energy than the average home of 1981.

Today an average of less than 1 out of every 6 public high school students is studying a foreign language, compared with almost 1 out of 4 in 1965.

An average of more than 1 out of every 2 households in America has a Sears credit card; 1 out of every 3 has a J.C. Penny credit card, and 1 out of every 4 has a Montgomery Ward card.

A European survey reveals that the average Englishman uses 40 ounces of soap each year.

Some kind of financial aid is received, on the average, by 3 out of 5 college students.

Each U.S. households consumes an average of 20 pounds of bananas a year.

An average of only 1 out of every 50 households in the U.S. does not buy bananas regularly.

An average of about 46% of the American women aged 18 to 34 plans to have 2 children; nearly 13% expects only 1; 11% expects no children. At this rate there would not be enough births to replace the current population and it would begin to decline.

The average housewife spends 25 hours a year making beds. She walks 10 miles a day doing chores around the house.

The average yearly cost to operate-

> a water heater is $215.17;
> a freezer (frostless-16.5 cubic feet), $92.82;
> a clothes dryer, $50.64;
> an air conditioner, $47.30;
> a range with self-cleaning oven, $37.23;
> a dishwasher, $18.51;
> a vacuum cleaner, $2.35;
> a hair dryer, $1.28;
> a clock, $.87; and
> a garbage disposal unit, $.36.

The average mobile-home price in 1980 was $15,500.

Of the 60,000 adolescent girls giving birth each year, an average of more than 9 out of every 10 decides to keep their babies.

The average American favors one of the following cookies: the chocolate-cream-filled sandwich, the vanilla-cream-filled sandwich, chocolate chip cookies, fruit-filled cookies, or vanilla-flavored wafers.

In 1978 Americans ate 2 billion pounds of cookies -- or an average of 26 pounds per household.

Sticky business-
The average tube of toothpaste in a family of 4 lasts 3 weeks.

Almost half of the nation's runaways average between 11 and 14 years of age.

It is estimated that by 1990 an average of 3 out of every 4 people in the U.S. will settle in coastal regions lying within 10 miles of the shore.

Not by bread alone–
An average English family eats 1 large loaf of bread per week; the French and Germans, $1\frac{1}{2}$; and the Italians – $2\frac{1}{4}$.

Federal regulations add an average 8 to 11¢ a pound to the cost of hamburger meat.

A poll in West Germany showed that an average of 1 out of every 5 German men brushed their teeth only on "special" occasions.

Every year, the average number of Muslims making a pilgrimage to the Great Mosque and the Kaaba in Saudi Arabia is 500,000.

By the end of 1980, at least 1 out of every 5 houses will have a microwave oven.

In the average year, 1 teenage girl in 50 gives birth to an illegitimate child.

Each year an average of over 1,000 new species of animals are discovered.

Salad again tonight?
Lettuce is by far the most important vegetable in America. Enough lettuce was eaten in 1979 for every man, woman, and child to have digested an average of 23.3 pounds of it.

An average of 7 out of every 10 American families expect children to earn allowances by doing household chores; 25% of these families withhold allowances as a form of punishment.

The average American prefers to turn on the news (60%) rather than read it (39%).

In 1972 the number of 1-family homes that had 2 or more baths averaged 5.3 out of every 10. By 1977 this figure had risen to 7 out of 10.

In a recent Gallup poll, an average of 94 out of every 100 claimed to believe in God; 50 out of every 100 claimed to believe that Adam and Eve were created to begin human life on earth.

In 1979 Protestants increased their church contributions from an average of $159.43 each to $176.33.

No dogs or kids allowed–
An average of 40% of 1-bedroom rental units and 20% of 2-bedroom units ban families with children.

Willy B. Handsome?
The most popular names registered on birth certificated in 1977 were Michael and Jennifer. Runners-up for boys: David, John, Joseph, Jason, Christopher, Anthony, Robert, James, and Daniel.
Girls best-sellers: Jessica, Nicole, Melissa, Elizabeth, Lisa, Danielle, Maria, and Christine.

On the average, each $1.00 a 6-member family spends on food, a 4-person group spends $1.03, a 2-person household spends $1.07, and a single person spends $1.11.

The average person ingests about a ton of food and drink each year.

The average American spends 53 minutes a day bathing and grooming.

Residents of Nevada bet an average of $846 a year in gambling casinos.

In September 1980, the U.S. Department of Labor reported that the average number of working wives in America was 1 out of 2.

More than 1 million woodburning stoves were sold in the U.S. in 1979, up 40% from 1978. Today 5 million American households are using woodburning stoves.

Head for the hills!
An average of 2 million teenagers run away from home each year. Over 55,000 of that number pour into Hollywood.

On the average, women are more likely than men to be living below the poverty line (59% women, 41% men).

Of the 160 million U.S. citizens of voting age, an average of 1 out of every 2 lives in the South or the West.

The National Association of Homebuilders claims that proposed government building standards which would save an average of 22% to 51% on energy would increase the average price of a new home by about $1,755.

Men die from an average of twice accidental deaths as women.

The cupboard is bare-
The average woman in America complains once a week of having nothing to wear.

Spilling the beans-
Of Americans aged 20 and over, 4 out of 5 drink coffee regularly.

Gallup's most comprehensive religion poll found that on the average,
> 94% of Americans believe in a God or universal spirit;
> 84% think the Ten Commandments remain valid today; and
> 70% think there is a "devil" or evil force.

According to Convenience Store News, an average of 8 out of every 10 purchases of cake, pies, doughnuts, mustard, and cookies are made on impulse.

Living quarters-
The number of households in which people live alone or with an unrelated person has jumped 66% in the past 10 years, or an average of 6.6% a year.

Going in style-
Newborn babies need diaper changes an average of 80 times per week.

Average consumption of fishery products in the U.S. was 12.8 pounds per person in 1977.

During an average lifetime, a person in the U.S. packs away 60,000 pounds of food - the equivalent of eating a blue whale stem to stern.

A man's castle-
In England, 3 out of 4 people live in 1 family houses. In the U.S., the average is 2 out of 3.

Until the time of the Caesars, the average person living in ancient Rome was a vegetarian.

Me and my toob—
The average American's favorite pastime is watching
TV. A distant second is reading, followed by going to
the movies.

According to the National Institute on Aging, the
average age of admission to a nursing home is 80 years.

Telephones in Japan average less than 40 per 100 households.

Telephones in West Germany average fewer than 35 per 100 households.

The average home garden is 595 square feet.

Open and shut case-
The average American eats 4% of his food standing with the refrigerator door open.

The average cost of the average home garden is $19; the average yield is worth $325.

School administrators reveal that on an average school day 6% to 11% of students play truant. Normal absence rate for illness is 4.5%.

I'll buy that!
An average of 61% of all supermarket purchases are made impulsively.

What a drip!
A dripping hot water faucet wastes an average of 40 kilowatt-hours of electricity per month – the equivalent of running a color TV 8 hours a day for 31 days.

Researchers recently claimed that, on the average, nearly 7 out of 10 children from broken homes are still unhappy 5 years after their parents' divorce.

On the average, students that are early risers get better grades than late risers.

Aw, Beans!
In the U.S. more than 100 million people drink an average of 3.2 cups of coffee per day.

An average Mexican home mortgage requires a 50% down payment, runs 10 years, and carries a 19.5% interest rate.

From 1970 to 1979, the number of people living alone in the U.S. increased by 6.4 million, an average of 706,000 a year.

A Zulu baby nurses at his mother's breast for an average of 19 months.

Who ordered the grey lump?
Beginning soon compressed and freeze-dried meals could be delivered to your home for an average cost of $2.50 per meal. This is the latest of an estimated 10,000 commercial spin-offs from the space program.

Who's got the funnies?
The typical American will spend an average of 2 years of his or her life reading newspapers.

The average number of legal abortions in the U.S. today is 27.4 per 100 live births.

Success story–
In 1870 an average of 1 out of every 50 Americans had graduated from high school. In 1975 that average was 3 out of 4.

Serving time–
The average yearly cost of keeping a student in college is less than keeping a prisoner in jail for the same period of time.

In 1900 a 5 pound bag of flour cost about 13¢; in 1975 it averaged $1. In the same period, a pound of sugar went from 7¢ to 37¢; a pound of round steak from 12¢ to $1.89; a pound of butter from 26¢ to $1.03. Now check the 1975 figures against what you are paying today.

The average American consumed 88.9 pounds of refined sugar, 263 eggs, 9 pounds of coffee, and .8 pounds of tea in 1975.

Children wolf down an average of 3 to 4 pounds of candy per week! The average adult American consumes 15 pounds of sugar per year.

The average American's vocabulary is comprised of 10,000 words.

The average presidential family has 2 sons.

The average American visits his or her doctor 4.9 times a year.

Buying a home? Are you 28 years old? Do you earn $20,000 a year? Are you a skilled worker or professional? Are you married but childless? If so, you are one of today's average home-buyers.

The average American diet includes 117 pounds of potatoes, 116 pounds of beef, 100 pounds of fresh vegetables, 82 pounds of fruit, and 283 eggs per year.

American consumption of apple juice increased an average of 74% a year from 1975 to 1980, increasing from 430 million pints to 800 million pints.

In 1978 23.5 million households took vitamins regularly, with an average annual bill of $55.32.

An average of 3 out of every 5 soda drinkers buy colas.

An average of 3 out of 5 Americans say they keep a household budget.

U.S. teenage boys eat an average of 1,917 pounds of food a year or 5.3 pounds a day.

In 1979 Americans spent an average of 7% more for daily newspapers, a total of $3.5 billion, and 15% more, $1.4 billion, for Sunday papers.

Mediacracy-
The average number of TV sets per 1,000 Americans is 571.

On the average, Americans in the higher income brackets eat less white bread than whole-wheat brands. Southerners prefer white bread, Easterners and Midwesterners prefer rye, and Westerners prefer French.

If inflation continues at the present rate, by the year 2000 today's average $60,000 home will cost $1,643,000.

Food advertisements cost the average family $28 a year.

An average size candy bar will cost $8.22 in the year 2000 at the present rate of inflation.

Seventh graders have the highest vandalism average of all students in the junior high or high school grades.

A Roper poll indicates that the average person looks forward to checking the mail, working on a hobby, and watching TV.

Average American teenagers contribute about an hour of their time to household chores daily.

A survey of 7,000 students found that 1,750 considered it all right to cheat on tests, an average of 1 out of every 4 students.

Deceptive selling practices cost the average American 12% of his annual spendable income.

Americans consume an average of 36 pounds of eggs per person per year.

Looking for a roommate?
On the average, elderly women are far more likely to live alone than elderly men: 42% of the women 65 years or older are living alone, compared to 17% of men the same age.

The average American spends less of his income (13.5%) on food than most other people. The people of India spend 59% and those in Sri Lanka spend 56.2%.

Don't unpack, kids!
On the average, Americans move every 5 years or 14 times in an average lifetime.

The average life span of a home freezer is 20 years; refrigerator, 15; and gas range, 13.

The average American consumes 16 pounds of cheese per year and 95 pounds of sugar.

Market value—
Of the average food dollar, 84¢ actually buys food. Packaging costs 13¢ and advertising and promotion the remaining 3¢.

The average Swede pays 20.2% of his income for mortgage or rent, while the Canadians spend 19.2%. Both the British and the Irish pay more than the 10.1% average Americans spend.

The average monthly federal cost for food-stamps in 1978 was $27 per month per participant.

The average 4-room new apartment in New York City in 1980 costs $1,600 per month, in Ireland, $263, and in Tokyo, $2,000.

On the bedpost overnight-
The average American will chew 175 sticks of gum this year, for a total of about 35 billion.

In 1979 Americans consumed about 250 million pounds of popcorn, an average of almost a pound for every person in the country. Most was popped and eaten at home.

The average first-grader in Russia receives an hour of homework each day.

On the average, less than 1 out of every 4 homes in Ireland has a refrigerator; in Spain about 1 out of 3 homes has one.

Diplomate-
The average U.S. citizen over 25 years old has completed 12.4 years of school.

The average American eats 4 slices of bread a day, less than half of the amount eaten 15 years ago. For every whole wheat slice, 3 white pieces are eaten.

In 1980 Americans ate, on the average, 41% more chicken than in 1978, 36% more greens, 35% more fresh fruits, 22% more cheese, and 21% more raw vegetables.

According to the Parent-Teacher Association, 9 out of every 10 high school students have had some experience with alcoholic beverages before entering high school.

Sole-power-
The average American owns 3.1 pairs of shoes.

The life insurance coverage carried by U.S. families averages $36,000 a family.

An average of 1 out of every 4 American adults misses breakfast regularly. Only 8% of people over 60 miss breakfast.

Americans eat an average of 245 pounds of meat per year - more than any other nationality, followed by Argentines and Australians.

The most popular last name in the U.S. is Smith, averaging 1 person out of every 2.4 million.

Americans who live in a metropolitan area pay an average of 54% more property taxes per person than rural residents.

Each American eats an average of 48.6 pounds of chicken a year.

Sugar Babies-
Children eat an estimated average of 3 to 4 pounds of sugar per week.

The average child under 10 years of age complains once a week that he or she has no one to play with.

I believe-
The number of Americans believing in life after death or some form of the hereafter averages 9 out of every 10.

The U.S. Labor Department reports that an urban middle-class family had to spend an average of $1,895 more in 1979 to maintain the same standard of living as the previous year, an average increase of 10.2%.

The average American consumes over 290 pounds of milk and cream yearly plus another 5 pounds of condensed and evaporated milk.

Worldwide the illiteracy rate is up. It has increased an average of 6.5 million people a year since 1970.

During the high-school years, the average teenager will spend over 600 days watching television.

Americans buy milk in gallon-size containers made of plastic an average of 64% of the time. Half-gallons sold in paper cartons sell to only 4% of milk-drinking America.

A survey of Public Opinion magazine indicates an average of 3 out of 4 Americans feel that raising a happy family is their most important goal.

From 1970 to 1979, average American consumption of butter droppped from 5½ to 4½ pounds per person per year; sugar fell from 102 to 96 pounds.

The average American ate 276 eggs in 1976, 36 less than in 1970, and drank 9 pounds of coffee instead of 14.

The average loaf of bread is more than a third water.

After purchasing a home and a car, the biggest expense the average person will incur is a funeral service, which averages well over $2,000.

An average of 4 out of every 5 housewives polled said the "no-stick" teflon used in frying pans sticks!

The average American house has a doormat.

About 1 out of every 2 doorbells does not work, on the average.

The average home furnace lasts 28 years without needing replacement.

How sweet it is!
American drink 300 12-ounce cans of soda per year. That equals 21.9 pounds of sugar.

Yankee go home–
Americans consume an average of 3,290 calories per day.

Uncanny–
Between 1976 and 1979, canned-fruit consumption dropped, but consumption of fresh fruit increased from an average of 81 pounds per person to 95.

Raising the roof!
The average price of a new house in the U.S. was $24,400 in 1970, $44,200 in 1976, and $78,000 in 1980.

Women visit the beauty parlor an average of twice a month.

Americans eat an average of about 18 billion hot dogs per year.

The big bowl game-
Each flush of the toilet consumes 7 gallons of water. Half of the average family's water is used in this way, an average of 39 gallons per person per day.

And baby makes 3.5-
The average American family has 1.5 children.

In the average year, there are an estimated 20,000 fires in mobile homes.

Worldwide there is an average of 9 phones per 100 people. The U.S. averages 95 per 100.

Stash the cash:
A Roper poll tells us than an average of nearly 3 out of every 4 Americans have regular savings accounts.

Alotta pencils-
The average annual cost per school pupil in the U.S. is $1,500.

Draw one!
Americans drink about 450 million cups of coffee a day, an average of $2\frac{1}{2}$ cups for every U.S. citizen over 10 years of age.

According to a Gallup poll, average Americans daydream about:

> travel 51%
> future life, such as being rich or smart 40%
>
> men's dreams - better job, being great athletes, having power
> women's dreams - future life, being smarter, being beautiful.

The average family of 4 needs to freeze 40 quarts of string beans, 40 pints of corn, 10 pints of peas, 15 pints of lima beans, and 20 pints of broccoli from their garden to have enough vegetables until spring.

The National Center for Health Statistics reports that aspirin deaths among children under 5 have been reduced by an average of 76% since safety caps were required on bottles.

A suburban mother spends 3 hours and 15 minutes of the average week "chauffeuring" her family.

According to the "Trivia Index," certain items are increasing in price much faster than most. Here are some average increases for the 1974-79 period:

McDonald's quarter-pounder with cheese 71%
postage stamps 88%
pack of M&M's 258%
shirts laundered 108%
band aids 51%
paperback books 112%

How sweet it is!
The British have the world's largest sweet tooth. They consume an average of 8 ounces of candy per person per week.

The average American eats $8\frac{1}{2}$ pounds of pickles a year. Dill pickles have an edge over sweet pickes 2 to 1.

The average Senegalese lives in the countryside, is uneducated, and lacks political awareness. He or she lives according to ancient customs, virtually unaffected by modern conveniences.

The average piece of meat is more than one-half water.

An average family spent 94% of its income on food, shelter, and clothing in the 1870s, compared to 62% today. They spent about 50% for food versus 25% today.

Hawaiians live an average of 4 years longer than other Americans.

Americans munch 400,000 tons of potato chips each year, an average of about $4\frac{1}{2}$ pounds per person.

Duck!
Worldwide, a commerical plane takes off or lands an average of every 3.5 seconds.

In 1980 the average family redeemed 70 newspaper coupons, each with an average value of 17.8¢.

Currently there are 456 million radios in the U.S., an average of 1.6 per person.

The average number of life insurance policies purchased by women leaped a whopping 28% between 1968 and 1978. This reflects the growing number of women entering the work force and the importance of their earnings to their families.

An average of about 1 out of every 4 Jewish homes has a menorah.

The average Frenchman uses 22.6 ounces of soap a year.

Total U.S. school expenditures in 1978 equaled $142.2 billion for 58.6 million pupils – an average of $2,400 per pupil.

About 4 out of every 5 12 to 17-year-olds read books daily. Girls average 80 minutes a day and boys less than 60 minutes.

4

The Good Life

Vintage port wine reaches maturity in an average of 40 years.

Bonkers for billiards--
During the 1960s, the average number of new billiard parlors built in Japan each year was 9,000.

There are 2 motion picture theatres in the Central African Republic, an average of 1 for every 4,100 people.

Joint custody-
The average number of people who have tried pot is 1 out of every 4 adults. Men are twice as likely as women to smoke it.

The average restaurant-goer favors to leave a tip rather than be charged an automatic 15% service charge, according to a recent Roper poll.

Down in smoke-
Americans are the world's heaviest smokers, averaging almost a pack per day per adult.

As the world churns-
Student-age men view an average of 1 soap opera a day, while women watch up to 4.

The average student soap-opera fan is 20 years old. About 40% watch because of intriguing plots, 30% because characters are so "dumb."

An average junk-food diet:

breakfast:	glazed donuts (2)
lunch:	hot dog
	sauerkraut
	mustard
	french fries
	soda
dinner:	sausage pizza
	pepperoni
	soda
snacks:	candy, potato chips, ice cream

The average American went to the movies 5 times last year, but the most fanatic moviegoers in the world are found on the island of Mauritius in the Indian Ocean, where the average per person is 19 times a year.

Between 1950 and 1980, average American consumption of soft drinks increased from 9.9 gallons to 33.7 gallons per person per year, while average beer consumption rose from 16.8 gallons to 22.6 gallons per person per year.

Is it 5 o'clock yet?
An average of more than 70% of all liquor is consumed in the home.

According to the Guinness Book of World Records, Australians guzzle an average of 52 gallons of beer per person per year.

Every glass of water served in a restaurant requires an average of another 2 to wash and rinse it.

In an average day, Americans are blitzed with 550 ads from radio, television, and billboards.

Trash flash!
Garbage can contents tell market researchers that people drink twice as much liquor on the average, as they admit to interviewers.

We're getting soft-
While alcoholism remains a problem, today's average American drinks only half as much hard liquor as the American of a century ago.

One ounce of LSD is potent enough to average 300,000 doses.

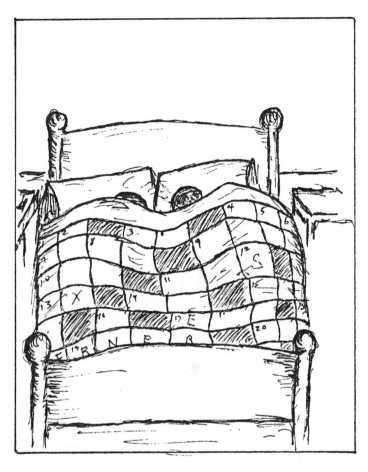

The most popular indoor game in the average household is doing.......the crossword puzzle.

During an average day, American restaurants serve 70 million meals.

TV newswatchers average 39 minutes a day; newspapaer readers, 17 minutes.

The average bingo player bets $5.93 per session.

The average bingo player is an 18-to-24-year-old female who is separated or divorced from her husband and has an annual income of $15,000.

Meals on wheels–
Americans today spend an average of 3 out of 10 of their food dollars eating out.

White wine sales averaged about 5 out of every 10 bottles of wine sold in 1979; red wine, about 3 out of every 10; and rose, about 2 out of every 10.

During the average day, Americans drink 60 million cans of beer.

Gamblers Anonymous estimates that there are 9 million pathological gamblers in the U.S. and this number is growing at an average of 10% a year.

An estimated average of $39 billion is bet illegally each year in the U.S.; about $20 billion is bet legally.

About 18 ounces of an average cola drink contains as much caffeine as 1 cup of coffee.

The average bill at an American-as-Apple-Pie luncheonette is $2.27, not including the average tip of 25¢. A typical lunch is a hamburger (medium), a cup of coffee with milk and sugar, and a dish of ice cream (chocolate).

The average movie has to gross $40 million to cover soaring production and marketing costs.

The amount of nicotine inhaled by the average pack-a-day smoker in 1 week (400 milligrams) would instantly kill a person if taken as a single dose.

Not to mention the worm—
Tequila ferments in an average period of $2\frac{1}{2}$ years.

Get the message?
Advertisers have spent an average of about $4 billion annually on network television during the last few years.

Between 1960 and 1976, the increasing popularity fast-food restaurants has resulted in the following rise in annual average consumption rates.

> Beef consumption rose from 64.3 to 95.4 pounds per person.
> Frozen potato use soared from 6.6 to 36.9 pounds per person.
> Catsup products rose from 7.6 to 13.3 pounds per person.
> Ice milk (used in cones and shakes) grew from 4.5 to 7.4 pounds per person
> Pickles went from 4.5 to 8.3 pounds per person.

Deep throat—
On the average, watchers of pornography, horror movies, and foreign films consume the most popcorn; sob-story fans buy the least.

On the average, 3 out of every 4 moviegoers are under 30 years of age.

A regular order of McDonald's french fries averages 50.5 fries per package.

The average smoker (1½ packs a day) smokes 10,951 cigarettes a year.

The average heavy smoker smokes as many as 30,000 cigarettes a year.

Chews your poison—
There is an average of 1 tobacco chewer for every 5 cigarette smokers in the U.S.

The District of Columbia led the nation in every category of alcoholic beverage consumption except beer drinking.

In 1979 4.3 trillion cigarettes were produced around the world, averaging about 1,000 for every member of the human race.

The best tips left to waitresses in restaurants are left by diners eating alone who use credit cards. They tend to tip an average of 23.7% of the bill.

Attractive waitresses got tips averaging 20.3% of the bill; less attractive waitresses get tips averaging 14.9% of the bill. Tips reflecting poor service average 11.9% of the bill.

In the U.S., an average of 3 out of 4 adults considers drunk driving a more serious problem than racial conflicts or pollution. It is considered our number 1 national problem.

The average high school senior has spent 15,000 hours watching TV and only 11,000 hours reading.

Cigarette smokers average 65% more colds than non-smokers.

Drunken driving is responsible for 25,000 deaths each year in the U.S., an average of 68.4 a day.

The average American over 12 years old eats at a fast-food restaurant 9.2 times monthly.

On the average, men dine out more often (9.7 visits per month) than women (8.6).

Show-stoppers
The average cost of producing a comedy or drama for Broadway in 1980 was $343,547; the average production cost of a Broadway muscial was $1,052,399!

The nation with the highest beer consumption is Belgium. The average Belgian enjoys 30.6 gallons of beer per person per year.

Studies show that on the average a person weighing 100 pounds will have 3 times as much alcohol in the bloodstream than a 200-pound person 1 hour after a drink.

The average American sees a dentist 1.5 times a year, though the American Dental Association recommends semi-annual checkups.

The average resident of Casablanca, Morocco, has never seen the movie Casablanca. The Humphrey Bogart film has never been shown there.

An average of more than 4 gallons of alcohol per year per adult are consumed by those living in Nome, Alaska.

The average chain smoker smokes more than 40,000 cigarettes a year.

Lacking many Western appliances, Soviet women spend an average of 35 hours a week on household chores.

The average American spends the equivalent of 404 days in the bathroom during his or her life.

Coffin nails-
Two-pack-a-day smokers live 8 years less than the average life expectancy; 1-packers live 3 years less.

The average American spends 24 years in bed during his or her lifetime; most of this time is spent sleeping.

The average American eats 8½ pounds of pickles a year.

The average Swede drinks more coffee than any other person in the world.

The average person ingests a ton of food and drink each year.

The average American eats 40 hot dogs a year. American eat enough franks each year to form a chain stretching from the earth to the moon and back.

The average Argentinian eats 10 ounces of meat a day.

Nevada residents bet an average of $846 a year in casinos.

The average cigarette smoker, smoking 1½ packs a day, smokes 10,950 cigarettes a year.

When eating in a restaurant, the average American will order chicken more often than any other entree on the menu.

The average Italian drinks 110.5 liters of wine per year. The average Frenchman is not far behind with 103.04 liters per year.

An American drinks an average of 255 calories per day in the form of alcoholic beverages.

The average Czech drinks 152.7 liters of beer each year; the average Austrian drinks 106.2 liters.

On the average, a sour pickle has 10 calories and an dill pickle has 11.

The average life span of the world's great musical composers is 60 years.

In an average game of Monopoly, players will land on Illinois Avenue more often than on any other space on board.

Consumers spend an average of 23% more when they shop with credit cards instead of cash.

A trade source estimates that the average annual wine consumption of Americans will be 4.3 gallons per person by 1990.

In 1978 Americans each consumed an average of 2 gallons of wine.

During Hollywood's heyday in 1939, film companies turned out an average of 2 films per day.

An average king-size cigarette will burn to its end in 7 minutes if lit but not smoked.

Potato chips average about 60% of all snack-food sales--beating out pretzels, popcorn, and corn twists.

5

Plants and Animals

Lions spend an average of 21 hours a day sleeping, catnapping, and staring into the distance.

On the average, lions stalking together catch prey almost twice as often as a lion hunting alone.

An electric eel can kill a person at an average distance of 15 feet away just by discharging an electric shock.

There is a breed of shark that averages 5 inches in length when fully grown.

The average life span of a goldfish is about 14 years.

On the average, 1000 bees must work their entire lifetimes to make 1 pound of honey.

The average length of a narwhal whale is 14 feet.

A baby blue whale drinks about 60 gallons of mother's milk a day and averages a daily weight gain of 200 pounds during the first 6 months of its life.

Americans spend an average of $3 billion on pet food each year.

The estimated average number of bats in the Carlsbad Caverns of New Mexico is 8 million.

An average beehive contains 50,000 cells.

The humpback whale's average annual migration is more than 4,000 miles.

It takes an average of 350,000 flower stigmas to make 1 pound of saffron.

The voracious shrew must eat on the average of every 2 hours or it will starve to death.

Dogs bite an average of 1 million Americans a year.

An elephant eats an average of 500 pounds of food each day.

That's a whale of a fish!
The humpback whale lunches on an average of 5,000 single fish.

Mayflies live an average of 6 hours. During that time they lay their eggs, which hatch 3 years later.

In an average year, swordfish contribute 10 million pounds of food for human consumption in the U.S.

Whale fisherman kill an average of 100 whales every day.

Greyhounds chasing a mechanical rabbit around a racetrack average speeds of 36 miles per hour.

The female flea lays an average of 400 eggs within 24 hours after feeding and mating.

The average housefly breeds in the center of the room; thus the instructions to hang flypaper away from corners.

Moving at a speed of 20 to 30 miles per hour, a flying fish can cover several hundred yards in the air. It actually glides on its winglike side fins.

King cobras account for 3 out of every 4 of the world's snakebite deaths, killing an average of 10,000 people a year in India alone.

The number of colorful songbirds is decreasing in the U.S. at an average of 1% a year due to the destruction of rain forests in Central and South America where they migrate.

A mature termite colony averages about 60,000 members; it eats only about 1/5 of an ounce of wood a day.

The head of an average-size whale, the largest creature on earth, weighs about 1200 pounds.

A single gram of king-cobra snake venom can kill an average of 150 people.

Start running–
A polar bear can smell a human at an average of 20 miles away.

The average flying lemur can glide through the air for a hundred yards, floating on the membranes between its front and back legs.

Terns (a type of sea gull) fly an average of 11,000 miles a year. They spend summer months in the Arctic and then migrate to the Antarctic in the fall, traveling halfway around the world twice each year.

It would require an average of 18 hummingbirds to tip the scale at 1 ounce.

Trained dogs have a better average success rate for finding gas leaks than any device yet invented.

The average worm is a great source of protein, containing 72% of the highest quality protein known to man.

3,000 over easy—
A queen bee lays an average of 3,000 eggs per sitting.

On the average, a shark's eyesight is 10 times keener than a human's.

Roses are red—
Americans buy more roses than any other flower. An average of 50 million are sold in New York City alone each year. The favorite color? Red!

The average electric eel can produce a shock measuring 600 volts.

Chow time!
Purina Dog Chow averaged sales of $27.5 million a month during 1980, making it America's top-selling pet food.

German shepherd dog bites are reported an average of 4 times more often than the bites of other breeds.

The average pig running at high speed can cover a mile in 7.5 minutes.

Oak trees are struck by lightning more often, on the average, than other trees.

A mouse averages about 500 heartbeats a minute; a human, about 80 an elephant, 20.

To satisfy its huge appetite, an elephant spends an average of 16 hours a day eating.

The average great white shark is afraid of the dark. Their vision is best suited to the light level of 60 feet underwater.

The average life span of a flea is 2.5 years.

Minnie Pearls-
The female oyster lays an average of 500 million eggs per year.

The American silkworm spins its cocoon out of a single thread averaging almost a mile long.

Not now honey, I have a headache-
During the time a female chimpanzee is in heat, she seeks sex an average of 20 times per day.

Look out below!
Young sparrows learn to fly just 2 weeks after their birth.

The average life expectancy of a gorilla is 60 years.

A full-grown oak tree will expel an average of 7 tons of water through its leaves in 1 day.

A bear cub averages 9 ounces at birth, although the mother may weigh over 500 pounds.

Oh rats!
Rats reproduce so fast that 1 pair plus its offspring can produce an average of 15,000 young rats a year; in 3 years that figure would be over 359 million rats.

In Korea the average number of snakes a day needed to prepare a soup called "baim tang" served in restaurants is 30,000.

What would a Camel walk a mile for?
Once the average camel gets moving, it travels along at a steady $2\frac{1}{2}$ miles per hour - as steady as a machine.

The average adult male Siberian tiger is over 40 inches high at the shoulder, 10 feet in length, and weighs 600 pounds.

On the average, an elephant's trunk can hold about $1\frac{1}{2}$ gallons of water.

Every 100 hens in the U.S. produced an average of 65.1 eggs a day in 1978.

In 1 hour of flying an average insect exerts so much energy that it may lose as much as a third of its total body weight.

Getting loaded-
When dehydrated, the camel drinks an average of 25 gallons of water in $2\frac{1}{2}$ minutes.

The average fruit bat or "flying fox" has a wingspan of over 4 feet.

The ai's (three-toed sloth) average speed is less than 1/10 mile per hour.

A whale's heart averages 8 beats per minute.

Busy as a beehive:
On a better than average day, a single beehive may gather 2 pounds of honey, the result of 5 million individual bee trips.

A whale's mating call is audible to other whales for an average distance of 100 miles.

If 1977 was your first year owning a pet, here is the average cost for the year:

    a horse, $2,900;
    a puppy, $375;
    a kitten, $200;
    a fish, $140;
    a hamster, $80;
    a parakeet, $62.

The female lion averages 90% of the hunting and killing, while the male grabs the first and best of the kill.

Water rises inside a plant's stem an average of 4 feet every hour.

On the average, more animals are killed by motorists than by hunters with guns.

The chicken population of the U.S. averages about 380 million.

The average Atlantic salmon can leap 15 feet high.

The average price of palm trees sold to Crown Prince Fahd of Saudi Arabia was $1,000 each. The Crown Prince bought over 100 to line the driveway of his palace. Coincidentally, they are called "royal" palms.

One dinosaur, the Branchiosaurus, averaged a weight of 62 tons. It was so heavy it could not stand up. It lived in the water because its legs could not support it.

The average life span of an African elephant is 70 years.

During an average year, more people are killed in Africa by crocodiles than by lions and other beasts.

Anty maim-
The giant anteater averages about 7 feet in length, of which 3 feet is its tail. It weighs almost 100 pounds, and its sticky tongue is 16 inches long.

Premium pets-
The average annual cost of pet health insurance is about $40 for a dog and $30 for a cat. This covers up to $350 a year in vet bills.

The life span of the grasshopper averages 5 months. They hatch in the spring and die before cold weather.

Leave it to beaver-
In the wild, beavers' dams have measured more than 1000 feet long. But on the average they pack it in at 300 feet long and $4\frac{1}{2}$ feet high.

The timing of the shrew-
Short-tailed shrews sleep only 1/12 of their lifetime, or an average of 2 hours per day.

The average vanilla bean comes from Madagascar. The island produces two-thirds of the world's vanilla.

Don't eat the daisies!
Houseplants are the main cause of children being poisoned in U.S. homes. The philodendron is the plant that most often poisons children.

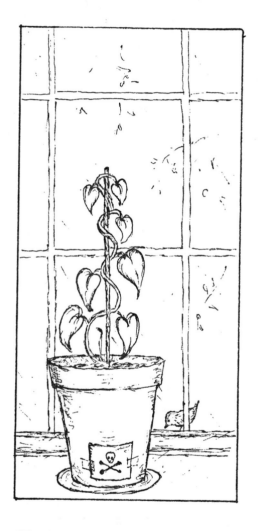

The honeybee kills more people with its sting during an average year than the most poisonous snake.

The walking fish of Asia can live for an average of a week out of water.

The 2-foot-tall alfalfa plant sends its roots down into the ground an average of 40 feet.

The average adult lobster is 8 years old. They grow to be about 13 inches long and may weigh as much as 9 pounds.

The female alligator lays an average of 40 eggs each time she has young.

On the average, cats are 6 times more prone to diabetes than dogs, owing to their high carbohydrate diet.

Polar bears in captivity eat an average of 29 pounds of food per day.

A bluefin tuna's swimming speed averages 60 miles per hour.

A shark swims an average of about 13 times faster than a strong swimmer.

The whale gestation period averages 360 days.

The average life span of a housefly is 15 days.

The Victorian water lily has leaves that average 6 feet across and that can support the weight of a large man.

The average porcupine has more than 30,000 quills.

Tarantula spiders can live an average of 2 years and 3 months without food.

The training of a seeing-eye dog requires an average time of 130 days.

The giant leatherneck species of turtle lives an average of 70 years and grows to abut 6½ feet long.

In the course of its evolution the elephant's trunk is getting longer at an average rate of about 1 inch every million years.

Fish story-
Can you name 2 species of fish whose average life span is longer than a man's? Here they are: the giant sea bass (75 years) and the sturgeon (over 75 years).

No surprise-
California produces an average of more turkeys per year than any other state, but Minnesota may pass it soon. Both of those states raise over 25,000,000 gobblers a year.

The average life span of a mayfly is 1 hour.

Americans keep 40 million birds as pets, an average of 1 per 7 people.

Australia's population records an average of 10 sheep per every human.

Mosquitoes are attracted to the color blue an average of twice as often as any other color.

Mush!
A dog-sled team at top speed averages 5 miles per hour.

Puppies eat an average of 35% more food when fed in groups than if fed separately.

In 1979 the average number of American homes growing indoor plants was more than 4 out of every 5.

The average Stegosaurus, a prehistoric creature, weighed about $6\frac{1}{2}$ tons, but its brain weighed less than 3 ounces.

An average of 12,500 puppies are born each hour in the U.S.

What a bouquet!
It takes an average of 2,000 pounds of real violets to make an ounce of violet oil, the "essential" oil needed to make violet perfume. But a chemist can produce 16 ounces of imitation violet oil from coal tar for about $3.00.

An average of nearly 1 out of every 2 people surveyed wears clothes or carries accessories with slogans, designer names, or other messages.

Eureeka!
An average of about 8,500 new species of insects are discovered each year by entomologists.

Tortoises living on the Galápagos Islands live to an average age of 170.

Endurance vile–
A hyena lives an average of about 25 years in the wild; a goat, 18 years; and a vampire bat, about 10 years.

The average life span of the termite is 30 years.

It is estimated rats eat an average of about 20% of the world's food crops each year.

On a hot summer day, a large oak tree gives off about 50 barrels of water in the form of a fine vapor.

On the average, a dog is more likely to bite a person in June than in any other month.

The average-size mature sequoia tree is about 300 feet high and 80 feet or more around at its base; yet the seed of the sequoia weighs about 1/7000th of an ounce.

Imagine the fish! Certain species of earthworms in Australia average between 6 and 10 feet long.

Your average-size aardvark weighs 150 pounds.

It costs an average of $2,500 to train a "hearing dog," a dog which responds and alerts its deaf master to sound.

The average life span of a black cherry tree is 200 years.

It's a bird - it's a plane...
The duck hawk has been timed in flight at 3 miles per minute, averaging 180 miles per hour.

The average-size porcupine has more than 30,000 quills.

A herd of 60 cows will produce an average of a ton of milk a day.

An average of more than 1 out of 2 female dogs in the U.S. are spayed.

Over easy, please-
The cod regularly deposits between 4 and 6 million eggs at a single spawning.

The average avocado contains 11 vitamins and 17 minerals.

Peregrine falcons average 175 miles per hour when swooping down on their prey.

Migrating ducks and geese average 60 miles per hour flying long distances.

A flying honeybee beats its wings an average of 250 times per second; a housefly, 190 times per second.

The average germinated seed contains Vitamin A, B-complex, C, D, E, G, K, and U and minerals such as calcium, phosphorous, potassium, and sodium.

King Konk–
The adult gorilla sleeps an average of 14 hours a day.

The whale generates the largest volume of seminal discharge, averaging 200 milliliters.

A shot of rye-
A single rye plant, after 4 months of growth, has 379 miles of roots, for an average root growth rate of 3 miles a single day.

Go fish!
The 9-inch archer fish can shoot a stream of water an average of 3 to 4 feet to knock its insect food out of the air.

The mature redwood tree grows to an average of about 300 feet high and lives about 1800 years.

Long live the queen-
The worker ants live an average of 7 years; the queen ants averages 15 years. Beetles live an average of 2 years; oriental cockroaches, 40 days; American cockroaches, 200 days; houseflies, 20 days; grasshoppers, 5 months; termites: worker, 20 years; queen: 50 years!

A snake's average speed is 2 miles per hour, except when frightened or aroused.

The beating of a mosquito's wings averages 1000 times per second.

The land snail moves at an average speed of 2 inches per minute.

The average poodle is more likely to bite a person than is the average sheepdog.

The average steer reaches sexual maturity 6 months after birth.

A squirrel stashes away an average of 20 bushels of food a season, although it only returns for about 2 bushels.

The common housefly can transport germs an average of 15 miles from the original source.

At birth the average baby elephant weighs 200 pounds. In an adult female's lifetime, she will produce 40 to 50 young, which stay with "the family" until their teens.

Animal hospitals are increasing in number an average of 3 times faster than human hospitals.

To each his own!
A snake can eat an animal 4 times the width of its own head in an average of 18 minutes.

The average human brain weighs 3 poounds compared to the sperm whale which has the largest brain in the world averaging 20 pounds.

It takes an average of 28 days to train a seeing-eye dog.

At top speed a sloth can average about 13 feet a minute along a tree branch and about half that on the ground.

A young sulphur-bottom whale grows to be an average length of 80 feet in 3 years.

Earthworms, eating an average of about 5 times their own weight a day, can turn municipal sludge and garbage into a nutrient-rich fertilizer.

An elephant's tusk averages 10 feet in length.

Lobsters weighing over 10 pounds were once common catches in U.S. coastal waters. Now the average lobster hauled up weighs 2.1 pounds.

The deer botfly has been clocked at speeds averaging 700 miles per hour.

The life span of a mouse is 5 years, on the average.

Americans spend an average of $170 million a year on bird feed.

A toad eats an average of 100 insects each day.

On the average, 20 newborn opossums can fit into 1 tablespoon.

The average "plant hospital" in the U.S. charges $58 to have a plant's disease diagnosed.

It takes an average of 71,000 elephants to supply 710 tons of ivory.

The spray of a skunk is so powerful that on the average you can smell only 0.07 ounce of it.

The fastest running mammal is the cheetah, an African cat that can run at an average of 45 miles per hour and hit 70 miles per hour for short distances.

While, on the average, dogs live longer than cats, the longest recorded life span of a cat is an amazing 34 years, against a dog's longest life span of 27 years.

The average basil herb repels flies and mosquitoes.

Only the average female mosquito sucks blood. She cannot reproduce until she has obtained it.

The average hummingbird weighs less than a penny.

The average honeycomb holds 4 pounds of honey.

An average ant can lift 50 times its weight. The average bee can manage 300 times its own weight - that's equivalent to a man pulling a 10-ton trailer.

6

The World Around Us

During an average year, California and Texas are the states with the most traffic fatalities.

In an average year, 49% of all forgeries are committed by women. The figure was closer to 14% 10 years ago.

Mexico's official murder rate averages about 46 homicides for each 100,000 people annually, the highest of any country.

The average IRS employee in the Taxpayer Assistance Program is trained for only 2 weeks. Yet the IRS operates on a manual that has over 40,000 pages.

Of Americans polled, an average of nearly 3 out of every 5 could not recall the names of the last 5 Presidents (Carter, Ford, Nixon, Johnson, and Kennedy).

On the average, nearly 1 out of every 20 runaway parents paying court-enforced child support is a woman, says the Department of Health and Human Services.

While Americans suffer with an average annual inflation rate of 13%, other countries are worse off: Argentina (140%), Israel (110%), Latin American countries (more than 30%), and African nations (more than 20%, except Egypt with 7.9%).

At an average of more than 2 to 1, diplomats said they feel serving in Washington is more dangerous than duty oversees because of the crime rate in the nation's capital.

FBI average statistics for 1979: 1 robbery every 68 seconds, 1 assault every 51 seconds, 1 car robbery every 29 seconds, and 1 burglary every 10 seconds in the U.S.

Federal and state regulations governing the production of the average hamburger number 40,000. They have been collected in a 311-page book.

Don't lock that door!
Police reports consistently show that more than half of all murder suspects had a close personal relationship to their victims.

The average U.S. President elected in a year ending in a zero does not live through his presidency.

The European nation with the highest reported murder rate is Luxembourg, with an average of 14.4 murders per 100,000 people. The lowest is Norway, with an average of 0.1 per 100,000.

In the 43 years since it opened, over 600 people have jumped to their deaths off the Golden Gate Bridge in San Francisco, an average of about 14 a year.

From 1975 to 1979, citizens of the U.S. Northeast and Midwest sent $165 billion more in taxes to Washington than they received in federal benefits--an average loss of more than $1600 for each taxpayer.

Most accidents in the average home are caused by bicycles, stairs and doors, in that order.

In 1979 the FBI reports that there was an average of 1 murder every 24 minutes in the U.S. and 1 rape every 7 minutes.

A violent crime is committed an average of every 28 seconds in the U.S.

This is a stick-up!
The average prison term served for robbery is 4.6 years.

An average of 1 out of every 3 people polled responded NOBODY when asked who is the most respected man in America today.

An average of 2 out of every 3 criminals released from prison will commit another crime.

The average state retail sales tax on personal property is 4%.

The National Taxpayers Union says that each American's share of the National debt amounts to about $112,910.

The cost of storing the 300 billion items of information that the U.S. keeps track of averages $500 per American, or $120 billion per year.

Peanut gallery-
In an average year, the U.S. government spends about $60 million to buy up the surplus peanut crop.

The U.S. government spends about $1.3 million a year to fund research programs to find ways of increasing peanut production.

Breaking away-
During Alcatraz' 29-year history, exactly 40 men made breaks, that's an average of about 1.5 a year. Most were recaptured within hours. Several were fished out of the water surrounding the island and a number were drowned in the raging currents. It is told that 1 man who was "drowned" survived and made his way to South America.

On average, 3 million people spent $1,363.44 per person on government in 1975; the average city of 50,000 to 200,000 people spent $765.84 per person.

The average shoplifter is a woman. Women are responsible for 85% of the country's shoplifting crimes. Most are amateurs and steal goods under $10.

More than half the people arrested for shoplifting are between 13 and 19 years old.

On the average, four times as many girls as boys shoplift.

In the 26 largest U.S. cities, an average of 1 out of 12 crimes is committed in the schools: 78% against students, 14% against non-teaching staff and visitors, and 8% against teachers.

The average U.S. President attended law school and/or practiced law sometime in his life.

The school crimes with the most frequent average of recurrence are assault and petty theft.

An average of 3 out of 4 car thieves who are convicted go to prison; the average sentence is 3 years.

The Federal Reserve System burns an average of $20 billion in worn-out bills each year.

From 1970 to 1980, the average yearly increase in the cost of running the White House, was 26.5%.

Burglars steal an average of $400 million a year from unlocked homes in the U.S.

A nationwide survey conducted by Research and Forecasts found that 5 out of every 10 Americans own firearms, that 4 out of 10 live in fear of being a victim in a rape, mugging, or assault, and that 6 out of 10 dress plainly to avoid attracting criminals' attention.

A United Nations agency estimates that the average ton of household garbage thrown out by a French family contains more than $250 worth of recyclable material, mostly aluminum.

Look both ways-
Men are twice as likely as women, on the average, to be involved in pedestrian accidents.

The average U.S. dollar bill has a life span of 18 months.

In an average year, 72,345 new patents are submitted for approval.

The old ways-
As late as 1970, the number of people in North Dakota that spoke German as their primary language averaged 1 out of every 8.

Conspicuous consumption-
The marquees of the 50 largest casinos and hotels in Las Vegas use enough electricity to run over a thousand average homes.

A house catches fire on the average of every 45 seconds throughout the U.S.

In the average, over 11 million crimes are committed against persons and property in the U.S. each year.

Look away, look away-
According to the FBI reports, the highest rates of murder and manslaughter occur in Alabama, Georgia, Louisiana, Mississippi, and Texas.

The number of library books circulated by the U.S. public libraries has grown from 543 million in 1950 to more than 1 billion in 1980, an average increase of more than 15 million per year.

An average of 4 out of 5 men serving prison terms for rape were convicted of statutory rape (having sex with an underage girl), not forcible rape.

An average of nearly 9 out of every 10 crimes committed in the U.S. involve theft.

About 10,000 art thefts are reported in the U.S. during an average year.

The average prison sentence is 39.7 months. Time actually served before parole averages half that.

U.S. Presidents receive an average of 3,500 letters from school children per week.

The poorest people in the world live in Rwanda, East Central Africa, where the average yearly per capita income is $80.

In 1979 the western states led the U.S. in inflation, with an average cost-of-living increase of 15.4% The north-central states also exceeded the national average increase of 13.3%, while the Northeast was lowest with 11.8%.

Most widely Red-
The two largest newspapers in the world are the Russian dailies Pravada (averaging 10 million copies a day) and Izvestia (averaging 8 million).

During 1980 the federal government spent an average of nearly a million dollars a minute, which is nearly a third of a billion per day!

An average of $22 billion will be spent in the next 10 years on alarm devices, especially for fire and burglary.

A University of Southern California study projects that 40 million microcomputers will have been sold to first-time users by 1990, an average of 1 for every 2 Americans households.

In 1975 Israel spent an average of $1,110 per person on military expenses.

An average of 70 people per day are shot with a handgun in the U.S.

The crime bill for the American taxpayer is increasing at an average of about 12% a year.

The average loanshark charges an interest rate of 5% a week or 260% a year. An FBI spokesman says loansharks have not raised interest rates despite 5 years of inflation.

The record industry loses an average of $125 million a month to pirates who make illegal copies for sale.

In 1979 California led the nation in bank robberies with an average of 166 a month, twice as many as second-ranked New York.

The 1979 national average for bank robberies was 120 per state.

According to an independent study conducted for the Internal Revenue Service, 26% of taxpayers failed to report some income. The average unreported amount was $530.

Senator James Sasser of Tennessee discovered that out of 241 trips by government personnel, 53% had been authorized by the person doing the traveling, averaging a little more than 1 out of every 2 trips.

The average number of Americans audited by the IRS averages 2.5 million per year.

Each day 8,200 homes in the U.S. are burglarized - an average of 1 every 10 seconds.

A dollar bill circulates for an average of 18 months.

Phoenix and Miami vie for the highest crime rates in the U.S. - averaging 9,482.2 crimes per 100,000 people each.

The Arab Palestinian guerrillas of al-Fatah get an average of $50 million a year from the Libyan government. President Qaddafi also gives financial and military aid to the Irish Republican Army fighting against the British.

Pinched-
An average of more than 1 out of every 2 people in the U.S. has received police tickets other than for illegal parking.

An average of 6 out of every 10 Americans carry credit cards. Half of the cards are from department stores. Next in popularity are oil company cards, Master Card, and Visa.

Line of duty-
Since 1965 an average of 94 law enforcement officers have been killed by criminals each year.

Oh give me a home-
In a recent year, the states of Washington, Wyoming, Utah, Montana, and Idaho led the nation in incidents of welfare fraud. Their combined rate of welfare fraud was $3\frac{1}{2}$ times the national average.

Doomsday room-
The average price of a fallout shelter is $3,995. It's more apt to be called a wine cellar or root cellar nowadays and has been known to house poker games.

Energy Pot-pourri-
Omni magazine reports that on the average more than 80 gallons of gasoline are used to produce 1 acre of corn; that lights produce up to 60% of the heat displaced by commerical air conditioners; and that supermarkets use 5% of all U.S. energy.

An average of 48.5% of the federal budget comes from individual income taxes.

The Glasgow Sunday Post is read by an average of about 4½ million Scots each week, about 77% of the reading-age population.

Despite the vast drain on human and natural resources, the United Nations reports that military spending around the world averages almost $1 million a minute.

The U.S. average murder rate is 48 times higher than those of West Germany, Japan, and Britain combined.

The U.S. spends a yearly average of $520 per American for defense. Among our allies, West Germany spends $396; France, $349; and Britain $314.

In the last 5 years, America's schools have reduced energy consumption by an average of 7% a year, saving taxpayers more than $1 billion in 1979 alone.

An average of more than 1 out of every 2 people released from prison are re-arrested within a 2-year period.

The typical grocery store shoplifter steals an average of $7.37 worth of goods per hit. About 1 out of every 3 grocery-store shoplifters are caught and prosecuted on the average.

Unclaimed tax refund checks totalling $150 million are giving the Internal Revenue Service a headache. The checks, all from the 1979 tax year, average $528. The checks are undeliverable due to faulty addresses.

Tax attack-
An average of more than 10% of tax returns filed by taxpayers earning over $50,000 are audited by the Internal Revenue Service, while only 2.6% of those under $50,000 a year are examined.

More than 4 out of 5 women earning $50,000 or more average a workweek of at least 51 hours. These top female earners average a 50% higher divorce rate than lower-paid women.

You can't keep a good man down-
The average male hospital in-patient in the U.S. is bedridden for 6 days; the average female, 8 days.

Since Franklin D. Roosevelt, U.S. Presidents have overrriden an average of 4.5 vetoes each year.

During an average year, more steel is used to make bottle caps than to manufacture automobile bodies in the U.S.

In Nicaragua there is an average of 30 homicides per 100,000 people every year. Spain has an average of 1 killing per million people every year.

An average of 69 people a day are shot to death with handguns in the U.S.

The world's population increases by an average of 90 million people every year.

The average Cuna Indian (living on the San Blas Islands off Panama's Caribbean coast) uses coconuts as currency.

7

Farming and Industry

In an average year, photocopy machines throughout the world produce 225 billion copies, and businesses spend about $14 billion on copier services.

The number of people joining the ranks of U.S. millionaires in 1979 was 316, reports the Internal Revenue Service, bringing the total to 2,092. Meanwhile the average income per tax return is $14,508.

The average number of photocopies produced each day in the U.S. is about 290.4 million.

Taxi drivers, streetcar conductors, and business executives have the highest average incidence of ulcers.

Between the years 1973 and 1978, American industry produced an average of 21% more goods annually while consuming only 1% more energy.

American farmers have an average of 1 tractor for every 88 acres of land under cultivation; Russian farmers have 1 for every 265 acres.

The Army averages fewer high-school graduates (40%) than the Navy (65%), Marines (67%), and Air Force (79%).

Pretty baby-
The average New York model's fee is $85 per hour.

Double-duty-
In 1976 the number of people holding 2 jobs in the U.S. reached an all-time high of 4.6 million; that's an average of about 1 out of every 50 people.

A survey shows that the average number of female business executives who began their careers as clerks is about 3 out of every 5.

In 1970 the Penn Central Railroad lost an average of $50,000 an hour, a total of more than $431,000,000. This is the worst business loss ever reported.

A postal worker's salary through rain, sleet, and snow averages $23,500 a year.

Between 1968 and 1979, female professors' salaries averaged $16,840 compared with $21,080 for men.

Company crooks-
In an average year, embezzlers steal twice as much money from banks as bank robbers.

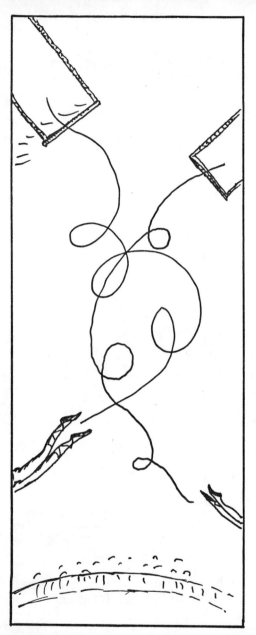

After deep-sea diving, the most hazardous jobs, according to an occupational study based on an average "extra deaths" per 1,000 per year, are those of astronaust (an extra 30 deaths per 1,000), speed boat and race car drivers (25 per 1,000), and trapeze and high-wire artists with no net (8 per 1,000).

On the hoof-
There is an average of about 2.5 million dairy and beef cattle for every state in the U.S. Several states have more cattle than people.

Wealthy, college-educated professionals are more likely than average to be the oldest child in the family.

With 28,000 members in Actors Equity and only 20% of them working, odds against an actor being employed in live theater average 4 to 1.

Swedish workers average the highest absentee rate among major industrial countries, Japanese workers average only 1.9 days.

The average supermarket stocks 12,341 items.

Scientists, engineers, and architects lie the least, on the average; politicians, actors, authors, and used-car salesmen lie the most, according to a study conducted by Dr. Robert Goldstein.

The average "take" for a pickpocket is $124 a grab; the average purse-snatching yields $98.

The average attorney's salary is $35,272; chief accountant, $34,865; engineer, $28,996; chemist, $27,881; and secretary, $13,612.

On the average, 7 out of every 10 computers in the world and 9 out of 10 electric typewriters are produced by IBM.

Time out!
Government workers average 15 days absence each, per year, due to illness; service industry workers average 14 sick days; blue-collar workers miss 12 days; white-collar employees miss 11; and farm workers, 10.

The American doctor works an average of 51 hours a week and takes an average of 4.8 weeks of vacation a year.

Out to lunch–
An average of 1,100 businesses per month flopped in the first half of 1980. This shows a 46% increase over the same period in 1979.

Don't count your chickens–
From 1920 to 1976, our farm population shrank an average of 428,571 people per year. In 1920 the farm population was 32 million people or 30% of the entire population. In 1976 the Bureau of the Census reported it to be 8 million or 3.9% of Americans.

Most modern automobile manufacturing plants produce cars on the average of 55 cars per hour.

The average wage of a chimney sweep in Southern California is $14.00 per hour.

The average unemployment benefit check nationwide is $100 per week.

The 10 largest U.S. industrial corporations have average assets of about $20 billion each.

An American doctor sees an average of 127 patients a week.

An average of about 11% of America's potato crop is used to make potato chips each year.

The average consumer complaint about cars and appliances involves the poor service of the item.

Pink slips-
A study of executives who had been fired shows that on the average (83% of the time) they did not aggressively seek their supervisor's attention for work well done.

Because of monotony, on the average of 58% of the naval officers who serve in the U.S. nuclear submarine fleet do not reenlist.

On the average, 1 out of every 5 business failures is caused by white-collar crimes committed by employees, according to Executive Digest.

An average of 1 out of every 6 American doctors is a woman.

Waiting room-
South Dakota has an average of only 96 doctors for every 100,000 citizens.

According to the Office of Consumer Affairs, an average of 7 out of every 10 customer complaints go unresolved. The worst offenders are the auto industry, domestic appliance manufacturers, the mail service, and clothing manufacturers.

Workers in Paris spend an average of 53 minutes in a restaurant for lunch; New Yorkers spend an average of 56 minutes.

The average number of American artists earning a living selling their own work is 2 out of every 50.

According to 68% of the secretaries polled, the average boss has handwriting that is either difficult or impossible to read.

P.S.-
The New York City Post Office handles an average of 20 million pieces of mail per hour.

In an average year, 3 out of every 4 Italians are involved in a labor dispute.

Let your fingers do the walking...
On an average day, a typist's fingers travel 12.5 miles.

Construction workers average the highest salary among blue-collar workers in the nation, $8 per hour. Lowest on the pay scale were service workers in restaurants, averaging $2.90 per hour.

The U.S. hotel industry estimates that working women average about 28% of all business trips a year.

The American industries with the lowest average success rate are: furniture manufacturers (85 failures per 100,000), transportation on equipment (76 failures per 100,000), textiles (73 failures per 100,000), and electrical machinery (72 failures per 100,000).

The average-size potato has no more calories than the average-size apple.

A Navajo weaver makes a 3x5 foot rug in an average of 16 days.

The U.S. Postal Service received an average of 17,028 complaints of mail fraud each month during 1980, a 27% increase over the previous year.

Americans work an average of 42.6 hours a week; Canadians, 38.9 hours; Swedes, 35.4 hours; Germans, 32.8 hours. Europeans average at least 2 weeks more vacation time per year than Americans.

A new study has shown that 1 out of every 2 prescriptions is so badly written that it can barely be understood by the pharmacists. Cardiac surgeons averaged best with 93% legibility, while general surgeons were worst with a score of 40%.

The Gross National Product worldwide in 1977 averaged $1,530 per person. The average in Africa was $400 per person; in Asia, $530; in North America, $7,020; in Latin America, $1,030; in Europe, $4,090; in USSR, $2,620, and in Oceania, which includes Australia and New Zealand, $4,490.

The average after-tax profit of leading American industries was 5½ cents per dollar of sales in 1978.

The average U.S. worker's wage rose 9.3% in 1980 but he is still losing money to inflation, which went up 13.4% over the same period.

About 53% of family physicians still claim to make house calls.

Thomas Edison averaged 26 new patents each year in his adult life.

A chemical, electrical, or petroleum engineer averages $25,000 salary the first year out of school.

A marble factory rolls out 1,000 marbles on the average of every 5 minutes.

The average American worker produces 25% more than the German worker and 65% more than his Japanese counterpart.

Take a memo-
If you're a secretary head for Peoria, Illinois where secretaries are paid 33% above the national average. Do not stop in Jacksonville, North Carolina, where salaries are 34% below average.

Upscale-
The average number of working musicians and composers in the U.S. that live in New York or California is 1 out of every 4.

In 1979 17.9% of all American workers were employed part-time. The average workweek of the part-timers was 19 hours.

South Korea's workweek averages $52\frac{1}{2}$ hours.

California averages 5 times more fishermen than any other state.

Mega-bucks-
An average of 1 out of every 2 of the 536 corporation heads interviewed in a survey claimed to be millionaires, and 1 out of 10 said he was worth over $5 million.

Growing 1 pound of grain requires an average of 100 pounds of rainwater.

The average price of opening a sun-tanning salon is $50,000.

Cab tab-
The medallion necessary to operate a New York City cab cost an average price of $42,000 in 1979.

Grain gain-
In 1800 it took an average of 400 man-hours of labor to raise 100 bushels of wheat. By the 1970's the same amount of wheat took only 10 man-hours to grow and harvest.

Trade deficit-
Annual exports from the U.S. average about $12 billion; imported goods average $15 billion.

Conscience money-
American business spends an average of $7 billion a year for pollution control. The federal government adds another $6 billion.

Don't forget the apples-
The average American teacher's salary is about $15,000 a year, and it increases about 6% a year.

To your health!
Since 1967 American doctors' fees have increased 230%, an average of 10% a year. During the same period, hospital room rates have increased about 15% a year.

Tillie the toiler-
Secretaries in New York or Toronto average 10 paid vacation days a year, while their European counterparts average from 15 (Dublin, Zurich, Vienna) to 30 (Madrid).

The average paperhanger covers about 10 feet of wall in an hour.

The average life expectancy of a new business in the U.S. is 6 years.

Suffer the little children-
A recent survey discovered that more than 1 out of every 3 teachers is miserable with the job and 41% would not become teachers again if they had their choice.

The average yearly financial loss to the consumer due to faulty merchandise is $142.

You're in good hands-
The 10 largest U.S. insurance companies have assets of about $20 billion each.

The businesses with the highest failure averages are: sporting goods (68 failures per 100,000 stores), men's wear (63 per 100,000), women's wear (60 per 100,000), and home furnishing (56 per 100,000).

It takes the average Russian worker 1 hour and 26 minutes to earn enough money to buy a dozen eggs.

The workers in China earn an average salary of about $40 a year.

Hitting the bricks-
The American worker who goes on strike stays out an average of about 2 weeks. Canadian strikers average about a month off the job.

Coincidence?
Using figures dating back as far as 2 centuries, one expert on long-range trends detects a rise-and-fall cycle of business failures, sunspot activity, combined stock prices, and gross abundance, averaging every 5.91 years.

Where the bucks are—
If you are a construction worker in New York, don't move; you're making the highest average salary in the business (over $13,000 a year).

There is 1 shoe-repair shop for an average of every 17,000 Americans.

Shoplifting costs every household in America an average of $200 per year.

TV directors of half-hour sitcoms make an average salary of $5,000 per episode aired.

TV directors of hour-long shows like "Dallas" average $10,000 per episode aired.

TV writers average the same salary as the directors—$10,000 per hour episode.

A 2 hour movie script for TV can be sold for an average of $35,600.

A TV director of photography averages $30.75 per hour or $246 per day.

A TV camera operator averages $16.58 per hour or $132.64 per day.

A TV Gaffer (chief electrician) averages $12.29 per hour or $98.32 per day.

A TV grip (equipment mover) averages $12.29 per hour or $98.32 per day.

A TV art director averages $18.46 per hour or $147.68 per day.

A TV costume designer averages $11.31 per hour or $90.48 per day.

A TV key makeup artist averages $12.29 per hour or $98.32 per day.

A TV sound effects editor averages $11.31 per hour or $90.48 per day.

The American farmer feeds an average of 5 times as many people as his Russian counterpart, with about half as much land and 1/6 as many workers.

An average of 2 out of 3 new job openings in the U.S. are in the Sunbelt and West.

In the U.S., New York, New Jersey, and Pennsylvania had the slowest average increase in job openings during the last decade (only 8.7%).

Personnel Journal found an 18% average improvement in mental work and a 17% average production increase on assembly lines when canned music was piped into offices and factories.

The production of steel from scrap metal requires an average of 75% less energy than the production of steel from newly-mined ore.

According to University of Michigan research, women work an average of 12% harder at their jobs than men.

A professional artist in America sells an average of 3.5 paintings a year.

The average American job claims 2 casualties per 1,000 workers per year.

On an automated chicken ranch each of 6 eating and watering periods lasts an average of 4 minutes.

Airline Executive magazine reports that airline captains who pilot international flights get an average salary of over $120,000 a year for flying about 75 hours a month.

Dig 16 tons and wattaya get?
Between 1973 and 1979, the U.S. non-agricultural worker's average weekly spendable earnings rose from $122.44 to $185.99. But his purchasing power, measured in 1967 dollars, dropped from $91.99 a week in 1973 to $85.77 in 1979.

There is an average of 1 doctor for every 535 people in the U.S. Washington, D.C. has the most per capita with an average of 524 per 100,000.

An average of more than 2 billion pencils are manufactured each year in the U.S. If these pencils were stretched out in a row, they would reach from the earth to the moon.

An average of more than 1 out of 2 U.S. residents now employed would like to continue working beyond the normal requirement age of 65, according to a recent poll.

Street vendors in California selling pretzels average sales of $10 a day on weekends.

California vendors selling leather goods average sales of $250 a day on weekends.

Belly dancers in San Francisco average $10 per hour but dance only 20 minutes of each hour.

More than 19 million families have 2 or more wage earners, producing a total average weekly income of $509.

The average male model today is 6 feet tall, a perfect size 40, and highly photogenic.

The average superstar male model makes between $175,000 and $200,000 per year.

Trash Flash!
Garbage can contents tell market researchers that the average person drinks twice as much liquor as he admits to interviewers.

Croupiers' salaries in Atlantic City, New Jersey casinos average $24,000 per year.

Count-y seat: the average purchase price of a seat on the New York Stock Exchange is $146,000.

Down on the farm—
The average American farm in 1850 consisted of 302 acres valued at $2,300.

The average American farm in 1975 had 387 acres and a value of $131,800.

The top 10 U.S. commercial banks have average assets of about $40 billion each.

Jolly beans—
On the average, only 100,000 pounds of rare Jamaica Blue Mountain coffee are produced each year.

Free enterprise—
Individual American taxpayers pay an average of 3 times as much income tax every year as U.S. corporations!

Even today 1 out of every 4 of the world's people averages an income of less than $200 a year.

The average yearly sales for a single House of Pancake restaurant tops 69,000 hotcakes.

Using tractors and bulldozers, Chinese farmers are dismantling an average of several miles of the 3,000-mile-long Great Wall every year. They use the stones to build huts and pig pens.

The big freeze-
In vegetable-freezing plants, it takes an average of 65 seconds to flash-freeze fresh vegetables for supermarket packaging.

On the average working women earn only 59¢ for every dollar men earn, so women work nearly 9 days to earn the same amount men earn in 5 days.

Washington, D.C., leads the nation in lawyers with an average of about 1,500 for every 100,000 people. New York is a distant second with 345 per 100,000.

An average of 1 out of every 3 cans of fish sold in America is consumed by a cat.

Deep-sea diving from oil rigs is today's most hazardous occupation, averaging a death rate of 1 out of every 100 each year.

A skilled tailor averages 45 stitches per minute.

Black American men and women average 35% of the enlisted strength of the Army.

Black women averaged more than 40% of the new female recruits in 1979.

Each year 68 million people are injured on the job, an average of 32 out of every 100.

In an average year, about 200 million marbles are made in 5 marble factories in West Virginia.

The jewelry industry uses an average of 1,000 tons of gold per year.

Dentistry uses an average of 87 tons of gold per year.

The industrial smoke produced in Britain in 1 year weighs an average of $2\frac{1}{2}$ million tons.

An average of 4 out of every 5 doctors in the Soviet Union are women. The average price of a facelift in Russia is $65.

There's big business, and there's BIG business.
The gross national product of the average country in the world is less than the total assets of American Telephone and Telegraph. With over $60 billion in assets, only 7 countries, including the U.S., had a higher GNP than A.T.& T.'s assets in 1973.

The average American holds 7.8 jobs during his working years, averaging 5.1 years per job.

The average American worker works 44 days per year to pay his taxes.

The average American man earns $675,000 during his lifetime.

The average American woman earns $347,739 during her lifetime.

The average use of land in New Zealand is to support livestock. New Zealand uses 70% of its land for its 60 million sheep and 2 million cows.

An average of 40,000 gallons of water flowed through ancient Rome's 9 aqueducts.

The average American man can cut and shuck 5 acres of corn in a season.

An average ton of manure, used as fertilizer, contains 10 pounds of nitrogen, 2 pounds of phosphorus, and 6 pounds of potassium.

Oats sprouted 5 days have an average of 500% more B-6, 600% more folic acid, 10% more B-1, and 1,350% more B-2 than unsprouted oats.

An average half cup of soybean sprouts has as much vitamin C as 6 glasses of orange juice.

The average amount of advertisers' and manufacturer's coupons is 14¢.

The average American uses 8 times as much fuel as the average person anywhere else in the world.

The average American family paid 11% of its income in taxes in 1953. In 1976 the income tax paid averaged 23%.

Most countries try to export goods to other countries, but Botswana, linked to South Africa, exports people. An average of 25% of the working population of Botswana works in neighboring South Africa's mines.

The U.S. Postal Service handles an average of 400 pieces of mail per person annually.

The average American household has a phone. Daily, there are more than 523 million phone conversations and 33 million long-distance calls made.

The average farm in Taiwan is 2.2 acres. By law, farms cannot be larger than 7.5 acres.

Poland is a Communist country and therefore, theoretically, ownership of businesses and agriculture should be collective. However, the average Polish farm is privately owned.

Nearly 85% of the farms in Poland were not collectivized due to the resistance of the peasants.

The average American family income in 1979 was $19,684, more than twice the average income of $7,933 in 1967.

The average American is a clerical, service, or sales worker. A total of 35,522,000 Americans work in these fields while 25,567,000 are professional or managerial workers.

An average acre of farmland needs 1,000 pounds of lime once every 5 years.

Wild walnuts will last an average of 2 years before drying up.

8

The Environment

In the year 1976, about 700,000 people were killed worldwide by earthquakes, an average of 58,333 per month.

During a snowstorm, the roof of a house collects an average of 20,000 pounds of snow.

The average number of flood days in Venice during the winter months is 24.

The Empire State Building has been struck by lightning an average of 23 times per year.

A tornado, spinning like a top, moves along at an average speed of 30 miles per hour.

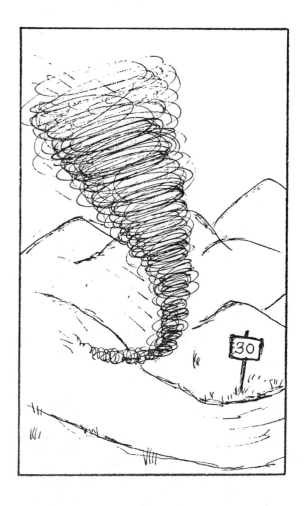

An average of 42 million square miles of the earth's surface is constantly covered by snow.

An average-size snowflake can be made with a million crystals.

An average-size snowflake takes 9 minutes to fall to earth from a height of 1,000 feet.

Each year the U.S. loses an average of 35 million acres of its woodland--altogether, an area about the size of Cuba. About 1 out of every 2 acres has vanished since 1950.

Each square acre of soil contains an average of 50,000 earthworms.

It takes an average of 4,500 gallons of water to grow the food in 1 person's typical daily menu of 2,570 calories; it takes an average 1,472 gallons to produce the typical fast-food menu of a hamburger, french fries, and a soft drink.

A tornado has an average life span of about 8 hours.

An average avalanche in the Swiss Alps may contain 5 million tons of snow and travel at 100 miles per hour. The wind it creates can blow down trees and houses half a mile away.

In 1 cubic mile of ocean water there is an average of 4 million tons of magnesium.

Going down?
The coast of Holland sinks an average of 1 foot each 100 years.

The world's population is growing at an average rate of nearly 1 million people every 5 days.

Contrary to all logic, the average raindrop is not tear-shaped. It is round or sometimes even doughnut-shaped.

The Washington Monument sinks an average of 6 inches annually.

The U.S. averages 1 forest fire every 30 seconds.

Zap!
Lightning strikes the earth 2 billion times a year, an average of 50 times a second.

The average number of people killed by lightning each year is 1 out of every million.

The U.S. fish catch alone averages 5 million tons annually.

Don't fence me in—
The city of Oslo, Norway, has an average of about 425 yards of space for every citizen; New York has about 25 square yards per person; the inhabitants of Tokyo have to settle for 1 square yard each.

Oh, the sun shines bright—
Louisville, Kentucky, averages only 93 sunny days per year.

The depth of glaciers averages 2,145 feet.

Iceland has a volcanic eruption an average of every 5 years. The water from the volcanic springs is used to heat the houses.

Sundown-
Each day the sun loses 360 million tons of mass, an average of 15 million tons per hour.

Alaska's 375.3 million acres cost the U.S. government $7.2 million dollars, an average of about 2¢ an acre.

L.A. means "lousy air"-
Los Angeles has the foulest air in the nation with an average of 318 days a year rated "unhealthful", "very unhealthful" or "hazardous."

Is Everybody happy?
A Nigerian study shows that of 4 out of every 5 people who live in rural areas most would like to move to the cities. In England, where 8 in 10 live in the cities, only 20% indicated they would like to remain in a city.

Average visibility while diving in the Alaskan waters is barely 15 feet.

On an average Texas summer morning, humidity reaches 91%.

The "sunbelt" cities of Miami, New Orleans, Birmingham, and Jacksonville lead the U.S. in average annual rainfall.

During the beginning of the century, Java, Indonesia, averaged 321 days of thunder per year. Figures show 3,200 thunderstorms per night!

An average cord of soft wood weighs 1 ton, a cord of hard wood weighs 1 3/4 tons.

It would take about 800 buckets of air to weigh as much as 1 bucket of water.

The average hurricane lives for only 10 days.

Lake Baikal in southern Siberia contains an average of 1/5 of the total world's supply of fresh fish.

A single lightning flash packs an average of 100 million volts.

The earth's crust measures 25 miles deep; it begins to get hot after an average depth of 5 miles.

To replace even half of America's oil needs with nuclear power by the year 2,000 would mean the construction of a large nuclear station on the average of every 3.5 days, according to <u>Omni</u> magazine.

An average of 1 out of every 7 acres of the earth's land surface is dry desert.

On a clear day, the average person can see $2\frac{1}{2}$ miles to the horizon.

Needle in a rockstack-
Diamond miners have to dig an average of 23 tons of ore to unearth 1 carat of diamond.

The average tornado will strike more often in the midwestern region of the U.S. than in any other region of the country.

It takes an average of 1 hour to spread enough salt on a 10-mile road to make driving passable.

The deep-
The average depth in feet of the Atlantic and Pacific Oceans is 12,998.

Santa Fe, New Mexico, receives an average of 17 more inches of snow than Fairbanks, Alaska.

The average rainfall in most of Niger is not enough to support any major agricultural ventures. So most of the people are subsistence farmers and very poor.

The U.S. government owns an average of more than 1 out of every 3 acres of land in the U.S., including 95% of Alaska and 86% of Nevada.

The average weight of a cubic foot of snow is 14 pounds.

An average of 4 out of every 5 Americans say they see no real gas shortage.

Of the 564,000 bridges in the U.S., 3 out of 4 were built before 1935. These bridges are 50 times more dangerous for motorists than regular roadways, which average 1 fatality for every 340 miles of driving.

The hottest ocean water in the world is found in the Persian Gulf, averaging 85 degrees Farenheit and sometimes reaching 96 degrees.

The average winter temperature in Reykjavik, Iceland, is higher than Chicago.

GASP!
In 1977 about 215 million tons of poisonous chemicals and particulates were emitted into the air in the U.S., averaging 588,888 tons a day.

During an average year, about 10 million tons of salt are spread on sidewalks, streets, and highways to melt snow and ice.

American industry and agriculture consume an average of 1700 gallons of water per day for every man, woman, and child in the country.

Snow job-
The average annual snowfall in Buffalo, New York, is 88.6 inches, the highest in the U.S.

Batting average-
An average of 75 tons of bat "droppings" were dug out of the Carlsbad Caverns in New Mexico every day, 6 months a year, for 15 years to be used as fertilizer.

Belgium averages 1 mile of railroad track for every 1½ square miles.

In an average year, 145 U.S. bridges collapse, and 29 more are ready to take the plunge at any time.

The wettest cities in Europe are Amsterdam and Brussels, where it rains an average of 206 days a year.

The average Soviet farm worker produces enough food for 5 people; his counterpart in the U.S. produces enough food for 60.

Between 1979 and 1980 the cost of diesel fuel shot up 77%. This affects food prices, because it takes an average of 7.4 gallons of diesel fuel to cultivate 1 acre of crops.

Approximately 1/3000th of the world's water supply evaporates yearly, on the average.

Every year an average of 2 million acres of land are converted to shopping centers, housing complexes, and other industrial projects in the U.S.

The average time elapsed between high and low tides on earth is 6 hours.

Americans save an average of 1 million barrels of imported oil a day and $7.7 billion a year in fuel costs by burning wood.

Arial view–
The 50 states of the U.S. average 46,270,000 acres each, but they range from vast Alaska (375.3 million acres) to Rhode Island (0.8 million acres).

Taking into account heating and pumping the water, the average American uses more energy shaving with a straight razor than an electric one.

In 1979, the number of earth tremors in the U.S. was 452, an average of 37.7 a month.

On the average, 1 cubic mile of sea water contains 166 million tons of dissolved minerals salts.

Flash!
Around the world lightning flashes an average of 100 times a second.

The average environmental noise level has increased by 100% every 10 years during this century.

On an average day, 1 billion gallons of petroleum is consumed worldwide.

The average land elevation of the earth is 2,700 feet above sea level.

In parts of Egypt the population density averages more than 250,000 people per square mile. Less than 4% of the nation's area is inhabited.

During the last Ice Age (18,000 years ago) the average temperature worldwide was only 4 degrees colder than today's average.

In the mist of things-
The foggiest place in the U.S. in terms of a single-year average is Cape Disappointment, Washington, which had 106 days of fog in 1979.

Gold Diggers-
South Africa leads the world in gold production, averaging about 700 tons per year, which is more than 50% of the world's annual production of 1,300 tons.

The Soviet Union produces an average of about 400 tons of gold per year; Canada averages about 49 tons; and the U.S. mines about 32 tons.

An average of 1 out of every 6 tons of the world's total supply of salt is used to control snow in the U.S.

9

The Universe

The Saturn V rocket carries enough fuel to drive an average car around the world 700 times.

The sun shines down through an average of 600 feet of water in the ocean.

Traveling in orbit around the earth, a spacecraft travels at an average speed 30 times faster than a jet plane, about 17,500 miles per hour.

The Voyager II spaceship hurtles through space at an average of 50,000 miles per hour.

On a clear night you can see an average of 2,000 stars.

And another one bites the dust-
On an average day, 1,000 tons of meteor dust hits the earth.

The earth and the moon travel around the sun at an average speed of 18.5 miles per second.

Meteorites in space have an average velocity of about 25 miles per second.

The average surface temperature of the moon is about 0 degrees Fahrenheit. During the lunar day, temperatures rise to at least 212 degrees Fahrenheit and during the night they drop to at least -238 Fahrenheit.

The distance traveled by light in a year is 6 trillion miles, averaging about 186,300 miles per second.

The average distance of the moon from earth is 238,859 miles. The farthest distance its orbit takes from the earth is 252,710 and its closest is 221,463.

Lost in the stars--
The average-size celestial body in our solar system is 201,240 miles in diameter. Most stars in our galaxy are 4 to 5 light years apart. Neighboring galaxies are probably about 1,000,000 light years apart.

Astronomers estimate that the universe averages 1 atom for 88 gallons of space.

The weight of our planet increases an average of about 10 tons every day, the result of a constant shower of fine space dust.

Astronomers calculate that the largest number of stars that an average person can see at 1 time with unaided vision is 3,000. With good field-glasses, about 120,000 can be seen and through a large research telescope about 1.5 billion stars can be seen.

The average speed at which the entire solar system circles within the Milky Way galaxy is 180 miles per second.

The entire Milky Way blasts through space at an average speed of 170 miles per second.

During a meteor shower, a single observer may see an average of 30 to 40 meteors every hour.

Spaced out-
An average of more than 4 out of every 5 people polled cannot remember the names of the 3 original American astronauts.

In an average month, earthlings can view 59% of the moon's surface.

A day on Uranus averages 11 hours in length earth time.

Bright but briefs-
The average duration of a meteor is 2.5 seconds, much longer than a lightning stroke 50 microseconds.

The average life of a planet is 10 billion years.

There is an average of 4,000 pieces of space junk still in orbit. Space junk pieces of rocket and debris which have broken away from the space ships may orbit for an average of a year before dropping toward earth and burning up in the earth's atmosphere.

The estimated life span of the sun is 10 billion years. Its present age is thought to be $4\frac{1}{2}$ billion years old.

10

Traveling

Of the 200,000 railroad crossings in the U.S. an average of only 1 in 4 has "gates."

The fastest train line in the world is the Japanese National Railroad's Hikari run, between Kyoto and Nagoya. The 83-mile trip is made in only 47 minutes, for an average speed of 106.5 miles per hour.

Today's average annual death rate for vehicle accidents is 4.5 people per 100 million miles. In 1901, when 3,150 people were killed in horse-drawn vehicle accidents, the death rate was 32 for every 100 million miles traveled.

On the average, more than 1 out of every 2 car-buyers in America buys cars that are 5 years old or older.

A car uses an average of 1.5 ounces of gas idling for 1 minute.

In the mid-19th century, the bicycle called a high-wheeler was widely used. The front wheel averaged 60 inches in diameter and the rear measured no more than 12 inches.

The Hertz Corporation reports that it costs more to operate a car in Los Angeles than any other U.S. city: an average of 44.2¢ a mile for a compact car, 12¢ more than the national average.

In 1980 the Newspaper Advertising Bureau found an average of about 9 out of every 12 new-car households had 2 or more cars, compared to about 8 out of 12, 4 years earlier. But they were putting less mileage on their cars, the survey found.

Prices up--travel down-
Because of skyrocketing European prices, travel in France is down an average of more than 20%.

Greece complains that tourism is off an average of 48% since 1978, and Spain and Sweden have dropped to 50% of their previous figures.

The American Automobile Association has declared the Chevy Malibu the closest thing to a "typical" car.

On the average, a stolen car is 200 times more likely to be involved in a crash than other cars.

The average American automobile horn beeps in the key of F.

Bad roads cost the average motorist more than $100 per year in tire wear, car repairs, and extra fuel.

An average of 2,332 motorcycles are registered per 100,000 people in the U.S. today.

Plug-a-bug-
Electric cars being tested today have an average range of 50 to 60 miles before recharging and take about 8 hours to recharge.

In 1979 10.9 million new cars were bought, averaging 908,333 per month; 9.2 million cars were sent to the junkyard that year, averaging 766,666 per month.

Researchers at the Massachusetts Institute of Technology (MIT) report that domestic airline jet passengers have a 1 to 2.6 million chance of dying on a given flight. The average risk of dying on an international flight is 4 times higher but still much lower than that of auto travel.

With a cloud of dust-
The average stagecoach in the 1860's could cover 112 miles every 24 hours, that's an average of nearly 4.7 miles per hour.

Stagecoach-less, a pony rider could cover an average of 250 miles a day in the 1860's.

The American motorist's average speed for a combination of highway and city driving is 40 miles per hour.

An average of 20% of all driving in the U.S. is done on interstate highways, even though they comprise less than 1% of the nation's entire traffic network.

An average of 51 cars a year overshoot and drive into the canals of Amsterdam.

The average American overseas traveler is a woman.

The average time it takes Amtrak's Metroliner to accelerate from 0 to 160 miles per hour is 2 minutes.

Barnacle build-
A steamship will collect over 100 tons of barnacles on its bottom during an average year.

Highway Magyar-
Hungary has the world's highest automobile death rate, averaging 112 each year for every 100,000 cars; Ireland ranks second with 87; the U.S. and Norway have the lowest, averaging 32 per 160,000.

Come-on-down!
The average American takes a warm-weather vacation.

To own and operate an economy car now averages 32¢ per mile. That is an average increase of 13.1% over the 1978-79 cars.

The 10 U.S. states that average the most travelers abroad are New York, New Jersey, Pennsylvania, Massachusetts, Illinois, Michigan, Ohio, California, Florida, and Texas.

Bring'em back, Charlie!
In 1980 General Motors recalled 1.3 million cars for defective seat belt anchors, yet an average of only 1 out of 3 has been brought in for repairs.

There are 150 million bicycles average 1 for every 2 cars on earth.

While the average number of drivers in Western states exceeding the 55-mile-per-hour speed limit is 2 out of every 3, the California Highway Patrolmen's Association says that its new cars will not go over 60 up hills and take 10 miles to crank up 92 miles per hour.

The average American will spend $203,164 buying and driving cars during his lifetime. The standard-size car that Mr. Average would have bought in 1957 for $2,000 will be a $17,500 compact in 2011.

In 1924 the average automobile cost $265.

In the 1980 Harris poll, an average of nearly 1 out of every 2 Americans planned to buy a small car, compared to only 28% who had expressed that intention a year before.

The average trip to ANYWHERE from Hawaii begins with a distance of 2,200 miles.

The average American vacationer is eager to visit new places but wants to travel with familiar faces.

A loaded freighter sails from New York to Los Angeles in an average of 14 days.

Schoolteacher John Marino shattered the California-to-New York bicycle record by cycling the continent in 12 days, 3 hours, and 41 minutes -- cutting 22 hours off the old record. John Marino averaged about 10 miles per hour for the entire distance.

Medical World News claims that an average of 60% to 70% of aggressive, high-speed male drivers have "macho" personalities: they are impulsive and belligerent. Such men are disproportionately represented among drivers in serious auto crashes.

The ocean liner Queen Elizabeth II with full passenger list averages only 12 passenger miles per gallon of fuel. More economical is the 12-coach diesel train, averaging 642 passenger-miles per gallon.

Almost 500 cars are stolen each year in the U.S. for every 100,000 Americans.

According to Goodyear Rubber Company, a person gets better average mileage-value out of tires than jogging shoes: Goodyear radial tires cost about $400 and last about 40,000 miles, averaging a cost of about 1¢ a mile. A $35 pair of running shoes lasts about 1,000 miles, averaging a cost of about $3\frac{1}{2}$¢ per mile.

During World War II the yards of American Shipbuilder Henry J. Kaiser turned out an average of 1 ship every 4 days.

Analyzing old records, the Hertz Corporation found that running the average car today is less expensive than it was 55 years ago.

A typical new car in 1925 cost $1,732, equal to $7,165 today.

Car depreciation was higher in 1925 than today, and maintenance costs, adjusted for inflation, were 3 times what we pay now. Operating expenses cost 50.3¢ a mile in today's dollars, compared to the present average of 40.1¢.

Gasaholics-
The 1979 Cadillac limousine and the Jaguar XJS average 10 miles per gallon of gasoline. They have the distinction of being the worst gas-guzzlers produced.

During the 1979 period, 215 million passengers traveled on Amtrak trains for a total of 4.9 billion passenger-miles. The average passenger trip was 22.8 miles.

The average American buys a used car instead of a new one. Used cars cost less and are less expensive to run.

The typical used car purchased in 1979 was a mid-sized sedan about 3 years old (showing 29,090 miles on the odometer) with power options and air conditioning, says a Hertz survey.

In 1968 Tibor Sarossy set a record for transcontinental motorcycle travel by riding from New York to Los Angeles in 45 hours and 41 minutes at an average speed of 58.7 miles per hour.

Van-tasy-
There are over 14 million pickups, vans, and panel trucks in use in the U.S. today, averaging about 1 for every 20 people. More than half are used for family fun.

A moped averages a savings of about 240 gallons of gas in the U.S. a year.

It costs airlines an average of $3 per in-flight meal served to coach-fare passengers. First-class averages $5 per meal.

On the average, bigger cars are safer in an accident, but studies show that smaller cars get into fewer smashups.

On the average, the foreign cars that have retained their best value for resale have been the Mercedes Benz 450 SL (75% value retained) and the Honda Accord (74%).

The worst car resale value goes to Mazda RX4, averaging only 37%.

An average mid-size car will go for $190,730 before taxes in the year 2000 at the present rate of inflation.

Lock your doors–
During the average year almost 1 million autos are stolen.

The average American spends 48 minutes commuting to and from work daily.

A machine called a shredder crushes junked cars in an average of 1 minute, breaking them into small, easily handled pieces.

In an experiment, students pretending to break into cars in New York City were challenged by onlookers an average of only 3 out of 100 times. Police confirm that car break-ins succeed 98% of the time.

Travel broadens one-
Spending a year in a foreign country is the average dream vacation. A close runner-up is taking an ocean liner around the world.

Bumper to bumper-
The average motorist drives 12,000 miles per year; 58% of the driving is city and 42% country.

The first auto race was held in Chicago in 1895. The contestants drove from Chicago to Evanston. The winner was Frank Duryea with an average speed of $7\frac{1}{2}$ miles per hour.

The average cost of renting an $8\frac{1}{2}$-ton World War II tank is $35 per hour.

The anticpated average cost of running an electric car marketed in 1984 will be 2.3¢ per mile (for the power only). A traditional gas-powered car getting 26 miles per gallon costs 5¢ per mile at 1980 prices.

An average of 1 million bicycle injuries and 1,000 bicycle deaths each year are due to collisions with cars.

Horatio Hornblower-
The average time that horn-honkers take to blast their horns at the driver in front of them after the light turns green is 8.6 seconds.

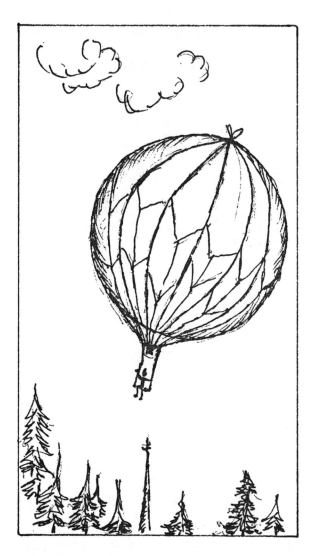

The 1980 price of the average stripped-down hot-air balloon is $6,500 (chase vehicle not included).

In 1975 approximately 1 out of every 2 people in the U.S. owned cars. In China, at the same time, 1 out of 5,632 had cars.

The New York subway system, with 230 miles of tracks and 462 stations, is the busiest in the world. It shuttles an average of 3 times as many passengers as the London system.

Day tripper–
The typical vacation spot averages 160 miles from home.

The American full-sized car that retains the best resale value is the Cadillac, with a resale value averaging 47.2% of its original value.

The Plymouth Gran Fury has the lowest resale value among American full-sized cars, averaging 25.9%.

According to Janet Guthrie, the race-car driver, if you want to keep the average American car running over 100,000 miles, you must:

> 1) give it a lube job on the average of every 6 months or 6,000 miles,
> 2) replace the ball joints every 35,000 miles or so,
> 3) repack the front wheel bearings on the average of 12 months or 12,000 miles, and
> 4) stick to the same brand of motor oil.

On the average, more than 4 out of 10 drivers have had at least 1 auto accident. Men are twice as likely to be in accidents as women.

Hurry up and waste–
At 50 miles per hour the average car gets a third more mileage per gallon than it would at 57 miles per hour.

By 1930 Indianapolis 500 drivers were averaging 100 miles per hour for the 500-mile race.

Car tombs-
On the average, your chances of surviving a car crash traveling at 75 miles per hour are only 1 in 40; at 65, 1 in 20; and at 55, 1 in 8, so says the National Safety Council.

The average resale value of a Chevrolet Corvette is 73% of its original value.

The worst American compact resale value goes to Pacer at an average of 38.7% of its original price tag.

Sports cars and subcompacts average a theft-rate 3 times that of other cars.

On the average, only 1 out of every 5 Americans buckles his seat belt in the car.

An estimated 1 in every 10 cars produced in the U.S. by 1990 will be electric-powered.

Saturdays are murder-
An average of 79 Americans are murdered and 191 die in car accidents each Saturday.

The average car trip is 5 miles.

In 1979 an average of 4 out of 5 business trips were taken by men.

U.S. airliners cruise at an average speed of 401 miles per hour.

Americans will buy 10 million bikes in 1981, an average of 1 for every 28 Americans.

The average American car in 1985 will be made mostly of plastic and aluminum, weigh a scant 1,300 pounds, and get 50 miles per gallon.

By the mid 1980's, the average U.S. car will contain 17% less glass and 18% less rubber than it now has.

The average driver can pocket at least $30 per year by using just 1/10 gallon less gasoline daily.

An average of 4 out of every 5 accidents do not occur in the rain or snow, but on dry roads.

The National Highway Safety Administration reports that an average of 3 out of 4 drivers surveyed favored the 55 miles per hour speed limit and only 10% are strongly opposed to it.

A manual-shifting car gets an average of 2 miles more per gallon of gas than an automatic-shifting car.

An average of 1 in 4 cars spot-checked was a quart or more short on oil; 1 out of 3 was low on coolant; 1 out of 2 had corroded battery terminals; 1 out of 3 was riding on 1 or more under-inflated tires; and 1 out of 2 had a dirty air filter.

From his study of 700 drivers, a German professor concludes that the average driver of red cars tends to be decisive; blue car drivers tend to be well-ordered and get fewer parking tickets; green car drivers are in tune with nature; yellow car drivers are looking for a change; white is chosen by fast drivers; gray is favored by boring people; and gold and silver by people who love money and brown by aggressive drivers.

The American Automobile Association considers sneezing a dangerous road hazzard. At highway speeds, the average 15 seconds of semi-control involved in a sneeze will make the driver almost blind for about 250 yards.

An average of about 8 million junk cars are recycled annually in the U.S.

The average cost of driving a typical car to work every day for a year is $3,176.

Watership down-
According to the U.S. Hydrographic Office, about 2,171 ships have been wrecked at sea over the past 100 years - averaging 21.7 ships a year.

Singapore's port has a ship arriving or departing an average of every 15 minutes.

The average passage-time through the Panama Canal is 8 hours.

The average American commutes 157,589 miles to work during his or her lifetime, the equivalent of travelling 6 times around the world.

The average American family owns at least 1 car. Over 28% own 2 or more vehicles.

11

Sports

Speed it up, Charlie!
A marathon runner can cover 26 miles in less than $2\frac{1}{2}$ hours, for an average speed of about 10 miles per hour.

In their 3 seasons of football at West Point, from 1944 to 1946, Doc Blanchard and Glen Davis scored 89 touchdowns and averaged 8.3 yards each time one of them carried the ball.

During their first 27 years in the majors, black baseball players led the National League in slugging averages for 22 years.

A line drive in baseball travels an average of 100 yards in 4 seconds. A fly hit to the outfield travels an average of 98 yards in 4.3 seconds.

Beattie Feathers averaged 8.5 yards a carry in 1934 for the Chicago Bears. This record was accomplished with 117 carries that took him 1,004 yards.

In 10 games, Steven Van Buren scored 15 touchdowns. This gives him the highest average of touchdowns per game for a season, 1.50.

Jim Turner holds the highest average for field goals per game in a season with 2.43 in 1968 for the New York Jets.

Ken Anderson completed 90.9% of his passes in a game on November 10, 1974, to hold the highest completion percentage record. His Cincinnatti Bengals were facing the Pittsburgh Steelers in that game.

Joe Namath hold the record for the most yards per game in a season. In 1967 the New York Jets superstar threw for more than 4,000 yards, giving him an average of 286.2 yards per game.

The Chicago Bears had an awesome defense in 1932. It held their opponents to an average of 3.1 points a game that year.

The 1950 Los Angeles Rams got the first down whenever they wanted it. In 12 games they averaged 23.2 a game for the National Football League record.

Baseball player Ty Cobb holds the record for the most hits in a career. With a total of 4,191, he averaged 174.6 hits per season.

Baseball player Cy Young won 507 games in his 22-year career. This gives him the amazing average of 23 victories a season.

The all-time greatest home run hitter Henry Aaron averaged 32.9 home runs per season. Babe Ruth averaged 32.6 home runs per season.

George Blanda averaged 77 points per season, scoring more points than any other football player. He made a total of 2,002 points.

Gale Sayers of the Chicago Bears is the all-time king of kick-off returns, with an average of 30.6 yards a return.

Football player Billy Johnson averaged 13.4 yards per punt return from 1974 to 1978, giving him the highest average ever in that column.

Speedy Duncan is one of the few football players to return more than 200 punts. He is sixteenth on the yardage-per-return charts with a 10.9 average.

Football player Sammy Baugh has the greatest punt average of all time, 45.1 yards per kick.

The great Jim Brown gained 15,459 yards in his football career. That gives him an average of 1717 yards a seson (including returns, receptions, and rushing).

Franco Harris has the most prolific record of all in post-season football. As of 1978, he had averaged 109 yards rushing and receiving in 14 post-season games.

The 1950 Los Angeles Rams averaged 38.8 points a game for the all-time National Football League record.

In 1933 Cincinnati Reds-yes, this is football-averaged 0.3 touchdowns per games.

David Lewis of the Cincinnati Bengals holds the American Football Conference record for the highest punting average. He averaged 46.2 yards a punt in 1970.

Football player Herb Adderly holds the highest lifetime average for yardage in interception returns. From 1961 to 1972, Adderly intercepted 48 passes and returned them for 1,046 yards. That gives him an average of 21.8 yards per return.

Football player Joe Golding holds the highest average for yardage in interception returns per season. In 1948 he averaged 51.3 yards for his 4 interception returns.

Walter Williams of the Chicago Bears picked off 2 passes and returned them for 148 yards on November 24, 1946. He holds the average record for a single game with a 74.0 average.

O.J. Simpson is a hard man to bring down on the football field. In 1973 he averaged 143.1 yards per game. The same year he rushed for 2,003 yards.

Football player Jim Brown holds the record for rushing attempts per game in a season. In 1959 his 290 attempts in 12 games gave him a 24.2 average. In 1977 Walter Payton tied that record with his 339 attempts in 14 games.

Cycling for 1 hour will burn up an average of 400 calories. Skipping rope uses an average of 300 calories an hour. Even something as inactive as standing at a cocktail party burns up to 20 calories an hour.

The average weight of a bowling ball is 13 pounds.

The average golfer has a 24 handicap.

The average jogger runs 1,438 miles per year, the equivalent of running from New York to Chicago and back.

The average major league batting average in 1978 was 24%.

Lionel Taylor averaged 56.7 catches per season in his 10-year career with the San Diego Chargers.

The average length of a world's heavyweight boxing champ's reign is 2 years.

An average ping-pong ball weighs .1 ounce.

Between 1867 and 1900, the Marquess of Ripon shot 371,000 game animals, an average of 216 a week.

Track experts estimate that the 100-yard dash requires, on the average, 95% speed and 5% stamina; the mile takes 50% speed, 50% stamina; the 26-mile marathon takes 5% speed, 95% stamina.

The average major league baseball player's salary in 1978 was in the ballpark of $99,000.

Golden oldies-
Billiards champions have the highest average age of any sport, 35.6 years.

In the average baseball game, batters hit 13 singles, 3 doubles, .5 triples, and 1.75 home runs.

The longest-sailing passing record in professional football is held by Sammy Baugh, who had a completion average of 70.3% in 1945, 128 completions in 182 attempts.

Sandy Koufax of the Brooklyn (later Los Angeles) Dodgers is the only pitcher to average more than 1 strike-out an inning during his entire career. Koufax struck out 2,396 batters in 2,325 innings.

Gimme an M!
The average home game attendance for University of Michigan football is 104,203. The stadium is packed at every game.

In 9 seasons of college and professional football, from 1957 to 1965, Jim Brown averaged 5.22 yards per carry and over 100 yards per game.

Ball grabber-
In the 8 National Football League seasons from 1962 through 1969, Lance Alworth averaged 5 receptions a game.

Football player Don Maynard's 11,834 yards gives him the record for most yards in a career. He averaged 18.7 yards a catch.

Derby dashers-
The fastest half-dozen Kentucky Derby winners averaged 2 minutes and 3/5 seconds for the mile-and-a-quarter run.

Bring a good book-
If you are following the action of a football game, you only have to look an average of about 20 percent of the time. Between-play activities consume the rest of the hour.

On the average, 1 out of 4 Americans went boating more than twice during 1978.

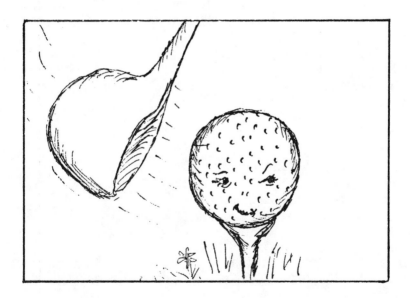

An average golf ball contains 336 dimples.

During the 17th century, a Japanese archer named Wada Daihachi shot 8,123 arrows down a 383 foot corridor over a period of 24 hours. That averaged 5 shots per minute.

Balls!

Fastball pitches average a speed of about 90 miles per hour, jai lai slams average 145, and a golfer's drive averages 155 miles per hour.

Basketball player Pete Maravich averaged 44 points a game during his career at Louisiana State University.

Mark Dyson, a junior horseshoe champion, won the 1980 U.S. title with 46 ringers out of 52 tosses, for a 91.9% average.

About 35 million Americans buy sport fishing licenses every year, an average of about 1 out of 8 people.

In a horse race, the "favorite" wins an average of 33% of the time.

The average recovery period of a football player following surgery for torn cartilage in the knee is 91 days.

On the average, third basemen playing in major leagues live longer than their teammates.

A table tennis ball weighs between 2.4 and 2.53 grams. That's 0.085 to 0.09 ounces.

Charlie Taylor of the Washington Redskins averaged 49.9 receptions annually, and he went on to catch more passes than any other receiver in the history of the National Football League with 649 receptions for 9.110 yards.

Your average American League pennant winner has to be a member of the New York Yankees; they have won it half the time for the last 60 years.

Hare racing-
The European hare flees its predators at an average speed of more than 40 miles an hour, taking leaps averaging 15 feet.

A racehorse averages a weight loss of between 15 and 25 pounds depending on the temperature of the air during every race.

Diamond glitter-
He had a .378 batting average for the season, smashed 59 home runs, and drove in 170 runs; but Babe Ruth did not win the Triple Crown in batting in 1921. The same season Ty Cobb batted .389 and Harry Heilmann batted .394.

More than 200,000 spectators turn out an average day at the Nurburgring auto race track in West Germany.

Floyd Baker, who played for the White sox in the 1940s and 1950s, hit an average of 1 home run every 2,280 times at bat during 12 major league seasons.

Legendary soccer star Pele received an estimated $4.7 million to play in 107 games for the New York Cosmos, an average of $43,925 per game.

Boxer Rocky Marciano won all 49 of his professional heavyweight bouts, with an 88% knockout average.

In his 25-year professional hockey career, Gordie Howe averaged 20 stitches in his face per year.

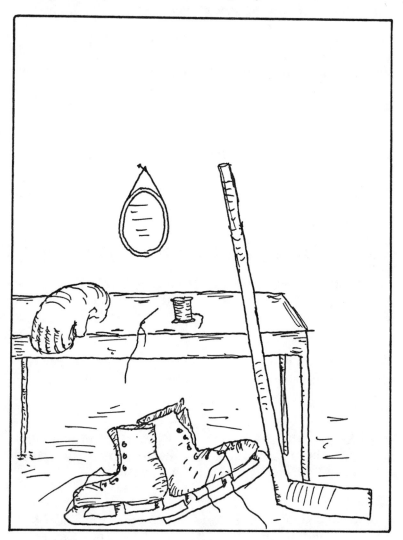

12

Miscellaneous

In 14th-century England, the number of males named Robert, William, Henry, John, or Richard averaged 2 out of every 3.

The glue on the back of a postage stamp averages 1/10 of a calorie.

An insurance company sample of drivers under 21 showed that those who had driver education average 12.8 claims per 100 cars, while those without it average 14.4 claims per 100. The average loss for those without driver training was about 11% higher.

The average modern elevator travels at 1200 feet per minute.

Birthright-
Maternity costs at the average metropolitan hospital are about $300 a day for the mother and $100 a day for the baby.

The average American dislikes using the Susan B. Anthony dollar. He feels it is too easily confused with a quarter and does not look like a dollar "should."

Vatican statistics show that the U.S. has an average of 120 priests for every 100,000 Catholics. Europe has 93 per 100,000; Africa, 33; and Latin America, 16.

Bogus bucks-
Check-bouncing went up an average of 41% in the U.S. in the past 4 years.

The Empire State Building's 6,500 windows are constantly washed in rotation. It takes an average of 15 days to get back to washing the same window.

Americans buy an average of $1 billion in waterbeds and waterbed equipment each year.

In super-high-rise buildings, like the World Trade Center in New York City, the high-speed elevators travel an average of 35,000 miles per year.

Traffic lights last an average of 18 years.

The average number of authors of each medical journal article is $4\frac{1}{2}$. Very few are written by a single physician.

It takes an average of 21 days for a grape to become a raisin in the California sun.

The number of Mormons in Utah averages 3 out of every 4 people.

The average wait in the New York Public Library for a requested book to be delivered to the pick-up desk is 36 minutes.

The right to bear arms—
The National Rifle Association estimates that Americans buy an average of 4 pistols over the counter every minute.

A fire hose with a $2\frac{1}{2}$-inch diameter delivers an average of 1 ton of water per minute.

The average wait for an ambulance to arrive on the scene of an acident in Washington, D.C., is 11 minutes and 34 seconds.

Bit sites—
More people are bitten by humans than rats during an average year. Hands and fingers received 61% of the bites and head and neck accounted for 15%, with the remainder spread over the rest of the body. The peak biting season seems to be spring and summer, according to the Assistant Health Commissioner of New York City.

The average life of a flourescent bulb is 30 weeks.

A new book is published in the U.S. an average of every 13 minutes.

An average of 15 minutes is needed to dig a grave using today's machinery.

More than 1 of every 6 men in Tibet are monks.

The average boiling time needed to make an ostrich egg hard-boiled is 4 hours.

The average pay telephone in the U.S. handle 18 phone calls per day. But a pay phone in New York City bus terminal averages 270 calls per day.

Out of every 14 checks issued in America, bounces. The average rubber check is for $19.47.

The average number of words in a long-distance phone call of 3 minutes duration is 350 words. That's if the callee doesn't interrupt.

The average refund check on 1979 federal taxes was $602.28. That is $103.23 more than it was in 1978.

An average "walk" takes 1 hour.

Nearly 4 million retired people received an average of $3,150 from pension plans in 1978.

The average total income for recipients of both Social Security and pensions was $11,440 for couples and $6,680 for singles in 1980.

In America an average of 4 out of 5 books are read by 12% of the population, and half the people have never read a book at all.

Greenland Eskimos chew an average of 1.75 pounds of gum a year, second only to Americans, who chew 2.2 pounds per year.

The average number of Americans polled who feel that advertising is "totally dishonest" was nearly 1 out of every 2.

The average American buying nonprescription drugs does not read the labels.

The average American burns 8 times more fuel than a citizen of any other part of the world.

The average sleeve length on a man's shirt is 34 inches.

An order of monks in Denver, Colorado, to keep their soup kitchen ladeling, sell a denim version of its own robe at an average price of $39.50.

The average person, according to a "Friendship" poll, feels that the most important quality of a friendship is the ability to keep confidences.

Action Comics from 1938, the ones you traded for baseball cards, are now worth an average of $7,000 each.

Picasso, the most prolific modern artist, averaged more than 5 separate works each week for a period of 75 years.

The famous .357 Magnum pistol propels a bullet an average of 1275 feet a second.

An average of 7% of America's water supply is lost annually due to leaky pipes.

The average light bulb burns for 62 days or 1,488 hours.

A study, on the average, finds that 9 out of 10 jurors reach a verdict when opposing attorneys make their opening statements and the trial itself seldom sways them

On an average day, 150,000 hospital beds remain empty.

More than 16 billion tiny toys have been packed in Cracker Jack candies in the past 68 years-an average of 235 million a year.

The average number of people who ride on elevators each year is 50 billion. Of all those ups and downs, there were only 1,459 accidents reported in 1979 and the majority of these were due to getting caught in closing doors.

The spice of life-
The average price for a pound of bay leaves is $183.83. Chives average $135 a pound and basil is about $1,172. Chili powder is the bargain at $9.11 per pound.

Soup is the last course served at the average Chinese meal.

Barbie Doll's manufacturers design an average of 36 new outfits per year for the doll and her friends.

The average number of Bibles distributed or sold worldwide every minute is 47.

The average length of a visit to a neighbor's house that is not considered a "rush" is 1 hour and 25 minutes in the U.S.

About 21 billion gallons of sewage are produced in the U.S. each day-an average of about 100 gallons per person!

Skywriters' messages last an average of 6 minutes before disappearing. On a calm day the message lasts longer.

Gabby garbage-
Based on a 6-year study of the garbage of Tuscon, Arizona residents, it was concluded that Americans throw away an average of 15% of the food they buy a year.

And you WILL lick it!
Postage for the average 1-ounce letter is 32¢ in Germany and 4¢ in Hong Kong.

The average number of letters lost in the mail is 1 out of every 265,000.

The average size of a prescription has grown 68% since 1960.

It cost the U.S. Census Bureau an average of $4.90 to count each American head in 1980, double the 1970 average cost.

An average of 3 billion Christmas cards are sent out annually in the U.S.

The average American does not know when his or her ancestors emigrated from their homelands to the U.S.

Americans who have had spiritual or mystical experiences during their lifetimes average 1 out of every 3 people.

The average time it takes to cure and hickory smoke bacon is 21 days.

Albert Tagora of New Jersey typed an average of 147 words a minute for 1 hour. During that stretch, Tagora ran off 8,820 words, which averaged $12\frac{1}{4}$ strokes per second.

College-educated Americans spend an average of 33 minutes a day reading the newspaper.

The average lead pencil will draw a line 34 miles long or write about 50,000 words.

The average reading speed for an adult is 300 words a minute.

Including John Hancock–
The 56 signers of the Declaration of Independence averaged 44.5 years of age at the time of the signing.

The actual aging period of Chinese "thousand year" eggs averages 41 days.

In 1880, when there were 48,000 telephones in the U.S., an average of 200,000 calls were made a day.

In 1975, there were 149,000,000 telephones in the U.S. and an average of 633,000,000 conversations were conducted each day.

To make a daguerrotype, an early photograph, required a 15-minute average exposure time.

Encounter-counter-
The Center for UFO Studies accepted over 30,000 reports of UFO sightings in 1978, an average of 657 per state. California led the nation with a total of 3,126 reports, including 403 reports "close encounters."

Partly sunny: Weather forecasts today average 75% accuracy.

Take a memo-
The average business letter in the U.S. costs $4.17 to prepare and mail.

In 1977, in New york City, diplomats attached to the United Nations received a total of 250,000 parking tickets, averaging about 684 a day. The representative from Guinea led the list with 516 tickets.

Supercalafradulisticexpedalidoshes-
The average American's vocabulary contains 10,000 words.

Be on the lookout!
The average number of UFO sightings per day is highest when Mars is closest to the earth.

Pit-stop-
The average deodorant user rolls on $1\frac{1}{2}$ ounces in 51 days.

Gone with the pin!
The average balloon pops in 3/1000 of a second.

The average American pushes a wheelbarrow when pulling is 46% easier.

Diamond Jim Brady's average breakfast as recorded by a New York restauranteur: a gallon of orange juice, 4 eggs, a quarter of a loaf of bread, steak with potatoes, grits, bacon, muffins, and several pancakes.

For his average everyday dinner, it was reported that Diamond Jim Brady used to order 36 oysters, 2 bowls of turtle soup, and 6 whole crabs as an appetizer.

To walk once around the Pentagon Building in Washington, D.C., takes an average of 17 minutes.

What does the average American think? According to a 1979 Harris poll:

> 53% of Americans feel there's been an increase in the number of doctors cheating on Medicare claims they make to the government;
> 51% feel that accountants embezzle money from companies.
> 50% believe that the practice of bribing politicians abroad by individual businessmen in America is on the rise;
> 48% feel that there has been an increase in businessmen cheating on expense accounts.

A semi-private hospital room cost about $189 a day in Alaska, the most expensive state, in 1980. Lowest was Mississippi at $65 a day, with a national average of about $110, according to the Health Insurance Association.

The average Dutch schoolchild does not know the story of the Dutch boy plugging a leak in a dike with his finger. That story was written by an American, Mary Mapes Dodge. However, the Dutch have built a statue to the story's hero.

The large "economy" size of an item usually is more economical, with an average savings of 15% on canned fruits and vegetables.

During the 1970s, the average yearly total of legal immigrants to the U.S. was 430,628, up 110,145 from the previous decade.

The average New York City building is 9.3 stories high.

In a study conducted by Dr. Robert Goldstein at New York University, it was found that if you are an Ameican you tell a lie an average of 1,000 times per year.

The average number of slaves per citizen in the Greek city-state of Sparta was 20 during its heyday in the 5th century B.C.

During the 1970s, the average yearly total of legal immigrants to the U.S. was 430,628, up 110,145 from the previous decade.

On the average, 7 out of every 10 Americans prefer to celebrate Independence Day on July 4 rather than on the first Monday in July.

In a crisis women tend to remain calmer than men on the average, researchers claim.

Not counting the replacing of dials and push buttons, the average life span of telephone equipment is 25 years.

The U.S. uses 1/5 of the cement manufactured in the world, an average of about 70 million tons per year.

Air conditioners completely recycle air throughout the Empire State Building on the average of every 10 minutes.

In 1979 the average American woman purchased 12.3 pairs of panty hose.

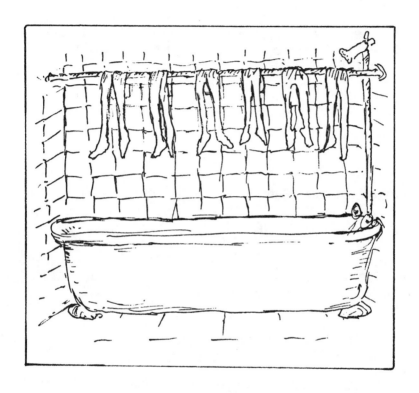

Getting anywhere?
On an average day you take 7,000 steps; you will walk 65,000 miles in your lifetime.

Ups and downs–
Each year an average of 8 miles of new escalators is installed in North America. This saves, approximately 62,000 steps and 43,916 feet of climbing.

Wolfgang Mozart averaged 17.7 original musical compositons per year during his brief 35-year lifetime.

It's weigh too little–
The average pencil-dot weighs 10 gammas (1/100,000th of a gram).

The National Endowment for the Arts' average individual fellowship grant is $8,500.

In a recent poll, 6 out of every 10 people felt that, on the average, store brands equal national brands in quality.

Of every 3 people contacted to participate in an opinion poll, 1, on the average, will refuse to do so.

The average elevator travels roughly 10,000 miles per year, the distance between New York and Melbourne, Australia.

There are 155 million telephones in the U.S.. averaging about 1 for every 1.8 people.

A recent survey revealed that the average number of adults in France who have not read a single book since childhood is more than 1 out of every 2.

During an average year, the yellow pages are consulted 17 billion times.